ADVANCED BDD OPTIMIZATION

# Advanced BDD Optimization

*by*

RÜDIGER EBENDT
*University of Bremen, Germany*

GÖRSCHWIN FEY
*University of Bremen, Germany*

and

ROLF DRECHSLER
*University of Bremen, Germany*

Springer

A C.I.P. Catalogue record for this book is available from the Library of Congress.

ISBN-10 0-387-25453-6 (HB) Springer, Dordrecht, Berlin, Heidelberg, New York
ISBN-10 0-387-25454-4 (e-book) Springer, Dordrecht, Berlin, Heidelberg, New York
ISBN-13 978-0-387-25453-1 (HB) Springer, Dordrecht, Berlin, Heidelberg, New York
ISBN-13 978-0-387-25454-8 (e-book) Springer, Dordrecht, Berlin, Heidelberg, New York

---

Published by Springer,
P.O. Box 17, 3300 AA Dordrecht, The Netherlands.

*Printed on acid-free paper*



Printed in the Netherlands.

# Contents

# Preface

VLSI CAD has greatly benefited from the use of reduced ordered *Binary Decision Diagrams* (BDDs) and the clausal representation as a problem of *Boolean Satisfiability* (SAT), e.g. in logic synthesis, verification or design-for-testability. In recent practical applications, BDDs are optimized with respect to new objective functions for design space exploration. The latest trends show a growing number of proposals to *fuse* the concepts of BDD and SAT.

This book gives a modern presentation of the established as well as of recent concepts. Latest results in BDD optimization are given, covering different aspects of paths in BDDs and the use of efficient lower bounds during optimization. The presented algorithms include Branch and Bound and the generic $A^*$-algorithm as efficient techniques to explore large search spaces.

The $A^*$-algorithm originates from *Artificial Intelligence* (AI), and the EDA community has been unaware of this concept for a long time. Recently, the $A^*$-algorithm has been introduced as a new paradigm to explore design spaces in VLSI CAD. Besides AI search techniques, the book also discusses the relation to another field of activity bordered to VLSI CAD and BDD optimization: the clausal representation as a SAT problem.

When regarding BDD optimization, mainly the minimization of diagram size was considered. The present book is the first to give a unified framework for the problem of BDD optimization and it presents the respective recent approaches. Moreover, the relation between BDD and SAT is studied in response to the questions that have emerged from the latest developments. This includes an analysis from a theoretical point of view as well as practical examples in formal equivalence checking.

This book closes the gap between theory and practice by transferring the latest theoretical insights into practical applications.

In this, a solid, thorough analysis of the theory is presented, which is completed by experimental studies. The basic concepts of new optimization goals and the relation between the two paradigms BDD and SAT have been known and understood for a short time, and they will have wide impact on further developments in the field.

## Acknowledgement

We would like to acknowledge the cooperation with Wolfgang Günther, Sherif Reda, and Alex Orailoglu. We would also like to thank Stefan Höhne for his help with the implementation of the techniques described in Section 5.2. Furthermore, we acknowledge valuable discussions, criticisms, and help from friends and colleagues.

Finally, we want to thank Mark de Jongh and Cindy Zitter from Kluwer Academic Publishers for their help with the preparation of the final manuscript.

Bremen, February 2005

Rüdiger Ebendt

ebendt@informatik.uni-bremen.de

Görschwin Fey

fey@informatik.uni-bremen.de

Rolf Drechsler

drechsle@informatik.uni-bremen.de

Chapter 1

# INTRODUCTION

During the last twenty years, the development of methods and tools for *Computer Aided Design* (CAD) of *Very Large Scale Integrated* (VLSI) circuits and systems has been a central issue in joint efforts of academia and industry. The interest in a further completion and improvement of *Electronic Design Automation* (EDA) will even grow in the future.

In the progress of this development it is important that computer-aided techniques keep up with the time. Facing the ever growing complexity of present-day circuits, these methods must still be able to provide a fault-free design.

The social relevance of EDA is obvious: fields of application vary from information and automation technology to electrical engineering, traffic engineering and medical technology.

Main problems of VLSI CAD are the automated synthesis of circuits and the optimization of circuits with respect to certain objective functions, e.g. chip area or operation frequency. Automated tools for these tasks have to accomplish the processing of very large amounts of data, often representable by so-called Boolean functions.

Already in 1938, Shannon proved that a two-valued Boolean algebra (whose members are most commonly denoted 0 and 1, or false and true) can describe the operation of two-valued electrical switching circuits. Today, Boolean algebra and Boolean functions are therefore used to model the digital logic of circuits where 0 and 1 represent the states of low and high voltage, respectively. To handle the large amounts of digital data arising during electronic design, efficient data structures and algorithms are necessary.

A well-known data structure for EDA is reduced ordered *Binary Decision Diagram* (BDD). BDD is a graph-based data structure for efficient representation and manipulation of Boolean functions and has been introduced in [Bry86]. It is known that BDDs allow a *unique* and very *compact* representation of Boolean functions. For these reasons, efficient algorithms for BDD manipulation do exist. Besides the clausal representation as a *Boolean Satisfiability* (SAT) problem, BDD is a state-of-the-art data structure for applications like equivalence checking, logic synthesis, formal verification of hardware, test and symbolic representation of state spaces in general, e.g. during model checking.

BDDs are well-known from hardware verification, e.g. from functional simulation, see [AM95, MMS$^+$95, SDB97]. BDDs are also used by techniques for logic synthesis [BNNSV97, FMM$^+$98, MSMSL99, MBM01, YC02] and for test. Here BDDs in particular support approaches to a synthesis for testability, see [Bec92, Bec98, DSF04].

All these applications benefit from *optimization of BDDs* with respect to different criteria, i.e. objective functions. In this, BDD optimizations happen at a deep and abstract logic level. In a design flow, this is an early stage before the step of technology mapping. After BDD optimization, it is often possible to directly transfer the achieved optimizations to the targeted digital circuits.

In the following, examples of BDD optimization and their applications are given.

- In BDD-based functional simulation, a simulator evaluates a given function by direct use of the representing BDD. A crucial point here is the time needed to evaluate a function. Hence, BDD optimizations have been proposed to minimize evaluation time [LWHL01, ISM03].

- One way to synthesize a circuit for a given function is to directly map the representing BDD into a multiplexor circuit. It is known that optimizations of the BDD (e.g. BDD size, expected or average path length in BDDs) directly transfer to the derived circuit (e.g. area and delay minimization). With that, the targeted applications are in the field of logic synthesis.

This book presents new work in the field of *BDD optimization.* Efficient search algorithms are presented to determine a good or optimal variable ordering of BDDs. The classical methods are based on different search paradigms, e.g. *hill climbing* and *Branch and Bound* (B&B). A recent suggestion is to use the generic $A^*$-algorithm. This is a fundamental search paradigm in *Artificial Intelligence* (AI). The use of this concept in EDA, however, is a recent innovation.

The classical criterion for the optimality of a BDD is its *size*, i.e. the number of nodes in the diagram. Classically, this criterion is addressed both with *exact* and *heuristic* approaches. After that, *alternative* criteria for BDD optimality are considered. Criteria different from BDD size have been studied in the classical works of [AM95, MMS+95] and in the more recent works of [SB00, LWHL01, FD02a, NMSB03, FSD04, EGD04b]. Applications for these alternative optimizations in logic synthesis, functional simulation and test are given.

The motivation for classical size-driven BDD optimization was the reduction of memory requirement and run time of the algorithms operating on BDDs. There has always been a strong demand for BDD size optimizations since BDDs are *sensitive* to variable ordering.

That is, dependent on the fixed order, in which input variables are tested along the paths from the root node to the terminal node, the size of the diagram may vary from *linear* to *exponential* [Bry86].

Hence, many approaches to determine at least a "good" ordering have been proposed in the past.

These methods can be roughly classified as:

- topology-based heuristics
- dynamic reordering
- simulation-based methods
- exact methods

The first class of methods are topology-based heuristics where structural information about a given circuit is used to determine a good ordering of inputs, e.g. see [FOH93]. Yet, these approaches are fast, but are often designed for classes of circuits with certain structural properties only and they do not guarantee an optimal result. In fact, experiments have shown that they can yield results up to 100 times the size of an optimal solution.

Second, dynamic reordering based on Rudell's sifting algorithm [Rud93] has become a widespread and very successful method. Here, a good ordering is determined dynamically by the system in situations where the application gets low on memory. This approach computes an ordering with what essentially is a so-called hill-climbing strategy, known from AI, e.g. [Ric88]. Since this normally happens fully automated and does not involve any inputs or actions from the user of such a software system, this method has become very popular. Today, many applications based on BDDs benefit from this widespread method. But still run time is an important issue and there is demand for faster solutions.

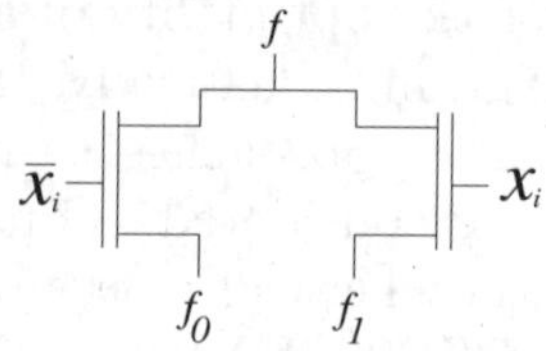


*Figure 1.1.* PTL multiplexor as wired OR of two NMOS transistors.

Further, simulation-based methods known from AI like *Evolutionary Algorithms* (EA) and *Simulated Annealing* (SA) have been proposed to obtain better results than the heuristic approaches before [BLW95, DGB96, DG97, GD00]. While these methods can obtain better results than mere "rules of thumb", the run time is usually much higher.

Finally, *exact* methods have been studied which are the only approaches that guarantee to determine an *optimal* variable ordering [FS90, ISY91, JKS93, DDG00, EGD03a, Ebe03, EGD03b, EGD04a, EGD05]. Except for the last two publications, these methods are all based on the B&B paradigm.

These classes of methods are given in increasing order of run time complexity. That is, exact BDD minimization is the hardest problem. In fact it has been shown that it is NP-complete to decide whether the number of nodes of a given BDD can be improved by variable reordering [BW96]. Yet there are applications in logic synthesis where a minimal number of nodes is needed since a reduction in the number of BDD nodes directly transfers to a smaller chip area: sub-optimal solutions can yield multiples of the minimum size and are not acceptable here. These applications follow multiplexor based design styles like *Pass Transistor Logic* (PTL). They allow to consider layout aspects during the synthesis step and by this guarantee high design quality. Recently, the interest in these approaches has been renewed, e.g. see [MBSV95, BNNSV97, $\text{FMM}^+98$, MSMSL99, MBM01, DG02, YC02].

Pass Transistor Logic uses only two transistors to realize a multiplexor (a wired OR of two MOS transistors, see Figure 1.1). However, both input polarities are needed to drive the pass transistor multiplexor. The advantages of PTL circuits as an alternative to static CMOS design are higher energy efficiency (low power), less circuit area and higher circuit speed. In 1995, Kuroda and Sakurai successfully used hand-designed PTL circuits to design digital systems [KS95]. A drawback of this early works however was the lack of automated synthesis tools. Later, several approaches for automatic PTL synthesis based on BDDs have been proposed [YSRS96, BNNSV97, CLAB98, SB00]. They all

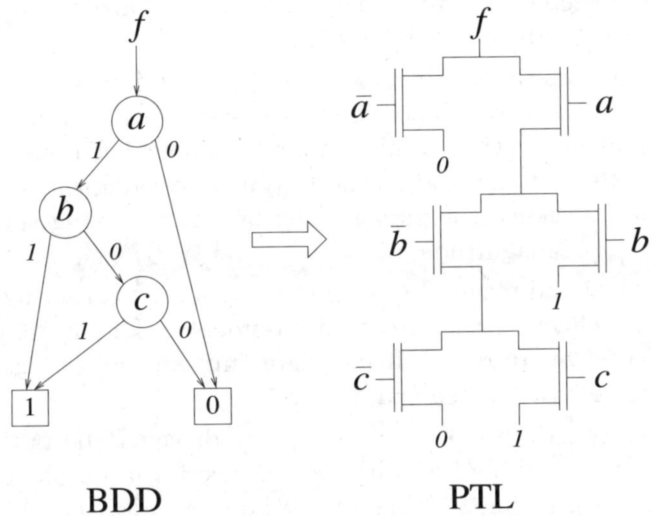


*Figure 1.2.* Mapping from a BDD to a PTL circuit.

benefit from the close correspondence of BDDs and PTL circuits which can be used for a straightforward mapping (see Figure 1.2).

A disadvantage of PTL designs is the high delay of transistor chains which is quadratic in the number of transistors used. Therefore, modern synthesis tools provide automatic buffer insertion and minimization [ZA01]. A second disadvantage, the possibility of *sneak paths*, i.e. connections of VDD and ground, already has been overcome with the use of BDD-based PTL synthesis since circuits derived from BDDs do *not* contain sneak paths.

This book starts the study of BDD optimization by presenting classical and the latest approaches to minimization of the number of nodes in BDDs. After basic notations and definitions are given in Chapter 2, algorithms are presented that exactly minimize BDD size in Chapter 3. The approaches vary from classical B&B methods to recent, more efficient approaches, which utilize extended B&B techniques. The latest development in the field is the shift to a new algorithmic paradigm, the generic *$A^*$-algorithm* [HNR68, Pea84, Ric88]. While this is one of the central concepts in AI, the EDA community has been more or less unaware of this paradigm for a long time. Recently, this approach has been suggested for exact size-driven BDD minimization [EGD04a, EGD05].

Besides presentation of the approaches, important algorithmic concepts and the theory behind them are introduced, such as heuristic state space search, ordered best-first search and the $A^*$-algorithm as well as the use of lower bounds during search.

All presented methods are evaluated by experimental results with benchmarks. In the case of recent approaches to exact BDD minimization, reductions in run time of more than 80% have been observed when compared to the best classical method. When applying the new method to arithmetic functions, the gain is even larger, achieving speed-ups of up to one order of magnitude.

Next, classical and recent approaches to dynamic reordering are given in Chapter 4, reflecting the latest developments. The latest promising method is based on Rudell's sifting algorithm and makes use of lower bounds to restrict the search space.

After these contributions to classical size-driven BDD optimization, in Chapter 5 the book focuses on *alternative* criteria for the optimality of BDDs. As a growing number of methods require the optimization of BDDs with respect to different objective functions, these new criteria become more and more important in comparison to the "classical" notion of BDD optimality, i.e. small or minimal BDD size.

The new criteria are related to *paths* in BDDs and they are motivated by applications in SAT solving, functional simulation, logic synthesis and synthesis for testability.

First, the minimization of the *number of paths* in BDDs is considered. The book gives a theoretical study as well as a minimization algorithm. The study is completed by experimental results.

Next, recent approaches to optimize BDDs with respect to the *expected path length* in BDDs are given and compared to previous approaches. The latest method is based upon Rudell's sifting algorithm. In contrast to the first approach of [LWHL01], the recent method is based on fully *local* operations, saving the high cost of touching large parts of the graph in every step. Experimental results with the new method are given, showing speed-ups of up to two orders of magnitude while preserving high quality of the results.

A *unifying view* is applied to compare the idea and the algorithmic hardness of the respective path minimization approaches. Then a new algorithm is derived, applying the unified view. The algorithm implements the minimization with respect to a new criterion for BDD optimality, the *average path length* in BDDs. This criterion, among others, has recently been suggested as a starting point for the synthesis for path delay fault testable circuits [DSF04]. The presentation of modern BDD optimization techniques is finished by giving experimental results for this last optimization approach.

State-of-the-art verification tools often use SAT solvers besides BDDs. Both SAT and BDD are well-established concepts. When looking at the experienced run times when using both paradigms for particular problem classes, the two concepts can behave orthogonally. The lastest trends in current and future research move towards the *fusion* of SAT and BDD.

In the book, the relation between SAT and BDD is studied in Chapter 6, considering formal equivalence checking as an example. This gives theoretical insight for a better understanding of the new trend and reflects the latest developments in the field.

Finally, concluding remarks are given in Chapter 7.

# Chapter 2

# PRELIMINARIES

In this chapter, some definitions and basic notations of Boolean algebra and reduced ordered *Binary Decision Diagrams* (BDDs) are given, as far as they are used in the following chapters. First, Boolean functions are defined and an important method for their decomposition is explained. Then a formal definition of BDDs is given.

## 2.1 Notation

This section explains general notations used throughout this book. Often sets and power sets are considered. The notation for the power set of a given set $M$ is

$$2^M = \{S | S \subseteq M\}.$$

$\mathbb{N}$ denotes the set of natural numbers *not* including zero, i.e.

$$\mathbb{N} = \{1, 2, \ldots\}.$$

Variables which are assumed to have values in $\mathbb{N}$ are most of the time denoted by the letters $i, j, k, m$ and $n$. Then, if the range of these variables is given by the context, sometimes a specification like "$n \in \mathbb{N}$" is omitted for simplicity.

When giving a result which expresses a both-way implication "if and only if" between a left and a right side of the statement, often the notation "iff" will be used as abbreviation, e.g.

$$a = \sqrt{b} \text{ iff } a^2 = b.$$

Functions usually are denoted using the identifiers $f$, $g$, and $h$. A function $f$ is given as a mapping from a domain $X$ to a co-domain $Y$. Domain and co-domain are stated in advance, e.g. $f\colon X \to Y$. The mapping then is defined with an expression of the form

$$f(x) = \text{an expression using } x$$

or

$$(x_1, x_2, \ldots, x_k) \mapsto \text{an expression using } x_1, x_2, \ldots, x_k$$

(e.g., in the case of a function $f\colon \mathbb{N}^2 \to \mathbb{N}$, the function may be given in the form $(x_1, x_2) \mapsto x_1 \cdot x_2$).

However, in the case of captions of figures etc. a shorter notation may be used to save space: the function above might be given in the short form

$$f = x_1 \cdot x_2.$$

This is done only if domain and co-domain are clear or given by the context (e.g. if $f$ above is already known to be a Boolean function, it is clear from the short notation that $f\colon \{0,1\}^2 \to \{0,1\}$).

## 2.2 Boolean Functions

Let $\mathbf{B}$:=$\{0, 1\}$ and $n \in \mathbb{N}$. Boolean variables, typically denoted by Latin letters, e.g. $x, y, z$ are bound to values in $\mathbf{B}$. Variables are referred to by subscripts which are from the set $\{1, 2, \ldots, n\}$, e.g.

$$x_1, x_2, \ldots, x_n.$$

To denote the set $\{x_1, x_2, \ldots, x_n\}$ of "standard" variables we use the notation $X_n$. Later, in Chapter 4, also the notation $X_j^i = \{x_i, x_{i+1}, \ldots, x_{j-1}, x_j\}$ will be used to refer to several subsets of $X_n$. The following is an introduction of notations defining Boolean functions.

DEFINITION 2.1 *Let $m, n \in \mathbb{N}$. A mapping*

$$f\colon \mathbf{B}^n \to \mathbf{B}^m$$

*is called a* Boolean function. *In the case of $m = 1$ we say $f$ is a* single-output function, *otherwise $f$ is called a* multi-output function.

These terms are used because Boolean functions are used to describe the digital logic of a *circuit.* A circuit transforms *inputs*, i.e. a vector of incoming Boolean signals to a vector of *outputs*, thereby following a certain logic. This logic can be described by a Boolean function.

Other properties of a circuit (e.g. critical path delay or area requirement) need a more sophisticated representation (e.g. as BDD which is a special form of a graph). Let $f\colon \mathbf{B}^n \to \mathbf{B}^m$ be a Boolean function. To put emphasis on the arity $n$ of $f$, we may choose to write $f^{(n)}$ instead of $f$. This notation will be used for functions in general, e.g. later on we use it to denote variable orderings $\pi^{(n)}$ (see Section 2.4). Sometimes we may even write $f^{(n,m)}$ to reflect the whole signature of a (Boolean) function $f$.

A multi-output function $f$: $\mathbf{B}^n \to \mathbf{B}^m$ can be interpreted as a family of $m$ single-output functions $(f_i^{(n)})_{1\leq i\leq m}$. The $f_i$'s are called *component functions.*

To achieve a standard, in this book the set of variables of a Boolean function $f^{(n)}$ will always be assumed to be $X_n$. If not stated otherwise, Boolean functions are assumed to be *total* (*completely specified*), i.e. there exists a defined function value for every vector of input variables. The Boolean functions constantly mapping every variable to 1 (to 0) are denoted one (zero), i.e.

$$\begin{aligned} \text{one}: \quad & \mathbf{B}^n \to \mathbf{B}^m; \quad (x_1, x_2, \ldots, x_n) \mapsto 1, \\ \text{zero}: \quad & \mathbf{B}^n \to \mathbf{B}^m; \quad (x_1, x_2, \ldots, x_n) \mapsto 0. \end{aligned}$$

A Boolean variable $x_i$ itself can be interpreted as a first example of a Boolean function

$$x_i\colon \mathbf{B}^n \to \mathbf{B};\ (a_1, a_2, \ldots, a_i, \ldots, a_n) \mapsto a_i.$$

This function is called the *projection function* for the $i$-th component.

DEFINITION 2.2 *The* complement *of a Boolean variable $x_i$ is given by the mapping*

$$\overline{x}_i\colon\ \mathbf{B}^n \to \mathbf{B};\ (a_1, a_2, \ldots, a_i, \ldots, a_n) \mapsto \overline{a}_i$$

*where $\overline{a}_i = 1$ iff $a_i = 0$.*

An interesting class of Boolean functions are (partially) symmetric functions. Later, in Chapter 3, algorithms for BDD minimization will be presented which exploit (partial) symmetry to reduce run time.

DEFINITION 2.3 *Let $f\colon \mathbf{B}^n \to \mathbf{B}^m$ be a multi-output function. Two variables $x_i$ and $x_j$ are called* symmetric, *iff*

$$\begin{aligned} & f(x_1, \ldots, x_i, \ldots, x_j, \ldots, x_n) \\ = \quad & f(x_1, \ldots, x_{i-1}, x_j, x_{i+1}, \ldots, x_{j-1}, x_i, x_{j+1}, \ldots, x_n). \end{aligned}$$

Symmetry is an equivalence relation which partitions the set $X_n$ into disjoint classes $S_1, \ldots, S_k$ called the *symmetry sets.* A function $f$ is called *partially symmetric,* iff it has at least one symmetry set $S$ with $|S| > 1$. If a function $f$ has only one symmetry set set $S = X_n$, then it is called *totally symmetric.*

## 2.3 Decomposition of Boolean Functions

DEFINITION 2.4 *Let* $f\colon \mathbf{B}^n \rightarrow \mathbf{B}^m$ *be a Boolean function. The* cofactor *of* $f$ *for* $x_i = c$ $(c \in \mathbf{B})$ *is the function* $f_{x_i=c}\colon \mathbf{B}^n \rightarrow \mathbf{B}^m$. *For all variables in* $X_n$ *it is defined as*

$$\begin{aligned} & f_{x_i=c}(x_1, x_2, \ldots, x_{i-1}, x_i, x_{i+1}, \ldots, x_n) \\ = \; & f(x_1, x_2, \ldots, x_{i-1}, c, x_{i+1}, \ldots, x_n). \end{aligned}$$

A cofactor of $f$ is the function derived from $f$ by fixing a variable of $f$ to a value in $\mathbf{B}$. Formally, a cofactor of $f^{(n)}$ has the same arity $n$. In contrast to all variables different from $x_i$, the variable $x_i$ is not free in the cofactor $f_{x_i=c}$. Hence the cofactor does not depend on this variable (see Definition 2.5).

Despite the generality of the last definition covering multi-output functions, sometimes only the cofactors of single-output functions $f\colon \mathbf{B}^n \rightarrow \mathbf{B}$ are of interest. When a multi-output function $f^{(n,m)} = (f_i^{(n)})_{1 \le i \le m}$ is given, we often consider the cofactors of the component functions $f_i$ only. These cofactors then are single-output functions of arity $n$. A cofactor of a multi-output function $f$ can be interpreted as a family of cofactors of the component functions of $f$.

A cofactor $f_{x_i=c}$ is sometimes called a *direct* cofactor to emphasize that there is only one variable bound to a value in $\mathbf{B}$. This opposes to a cofactor in more than one variable. E.g., for $k \le n$, $x_{i_1}, \ldots, x_{i_k} \in X_n$ and $c_1, \ldots, c_k \in \mathbf{B}$, the function $f_{x_{i_1}=c_1, x_{i_2}=c_2, \ldots, x_{i_k}=c_k}$ is a cofactor in multiple variables. This cofactor is equivalent to several direct cofactors, e.g. to

$$(f_{x_{i_1}=c_1, x_{i_2}=c_2, \ldots, x_{i_{k-1}}=c_{k-1}})_{x_{i_k}=c_k}.$$

In general it is equivalent to

$$(f_{x_{i_1}=c_1, x_{i_2}=c_2, \ldots, x_{i_{j-1}}=c_{j-1}, x_{i_{j+1}}=c_{j+1}, \ldots, x_{i_k}=c_k})_{x_{i_j}=c_j}$$

for any $1 \le j \le k$. A cofactor in multiple variables is uniquely determined regardless of the order in which we fix these variables. Hence, these cofactors can also be thought of being obtained by simultaneously fixing all the involved variables. To obtain increased readability, sometimes a "|" sign is used to separate the function symbol from the list of variable bindings, e.g. we write $f_j|_{x_{i_1}=c_1}$ for a cofactor in a component function $f_j$.

DEFINITION 2.5 *Let* $f\colon \mathbf{B}^n \rightarrow \mathbf{B}^m$ *be a Boolean function and let* $x_i \in X_n$. *Then function* $f$ *is said to* essentially depend *on* $x_i$ *iff*

$$f_{x_i=0} \neq f_{x_i=1}.$$

*The set of variables which $f$ essentially depends on, is called the* support *of $f$ and is denoted* $\text{support}_{n,m}(f)$. *Usually $n, m$ are clear from the context and hence we simply write* $\text{support}(f)$.

Formally, $\text{support}_{n,m}$ is a mapping $\{f \mid f\colon \mathbf{B}^n \to \mathbf{B}^m\} \to 2^{X_n}$. If not stated otherwise, a given Boolean function $f\colon \mathbf{B}^n \to \mathbf{B}^m$ is always defined over the variable set $X_n$ and always $\text{support}(f) = X_n$ is assumed. The next definition characterizes cofactors of a Boolean function which are derived by fixing at least one variable of the support of the function.

DEFINITION 2.6 *Let $f\colon \mathbf{B}^n \to \mathbf{B}^m$ be a Boolean (single or multi-output) function over $X_n$. A cofactor $f_{x_{i_1}=a_1, x_{i_2}=a_2, \ldots, x_{i_k}=a_k}$ with $x_{i_1}, x_{i_2}, \ldots, x_{i_k} \in X_n$, $(a_1, a_2, \ldots, a_k) \in \mathbf{B}^k$ is called* true *iff*

$$\{x_{i_1}, x_{i_2}, \ldots, x_{i_k}\} \cap \text{support}(f) \neq \emptyset.$$

Note that true cofactors of a function cannot be equivalent to the function itself.

Often it is more useful to consider a set of cofactors of a Boolean function rather than considering just one particular cofactor. This is reflected by the next definition.

DEFINITION 2.7 *Let $f\colon \mathbf{B}^n \to \mathbf{B}^m$; $f = (f_i^{(n)})_{1 \leq i \leq m}$ be a Boolean multi-output function essentially depending on all its input variables and let $I \subseteq X_n$. The set of non-constant cofactors of $f$ with respect to the variables in $I$ is denoted* $\text{cof}_{n,m}(f, I)$. *Formally, a function* $\text{cof}_{n,m}$ *is given as*

$$\text{cof}_{n,m}\colon \{f \mid f\colon \mathbf{B}^n \to \mathbf{B}^m\} \times 2^{X_n} \to 2^{\{f \mid f\colon \mathbf{B}^n \to \mathbf{B}^m\}};$$

$$(f, \{x_{i_1}, \ldots, x_{i_k}\}) \mapsto$$
$$\{f_i|_{x_{i_1}=a_1, \ldots, x_{i_k}=a_k} \text{ non-constant } \mid 1 \leq i \leq m, (a_1, \ldots, a_k) \in \mathbf{B}^k\}$$

*for* $1 \leq k \leq n$.

The set $\text{cof}_{n,m}(f, I)$ is the set of all distinct (non-constant and single-output) cofactors of $f$ ($f$ is interpreted as a family of $m$ $n$-ary single-output functions) with respect to all variables in $I$. Note that this is not a multiset, hence functionally equivalent cofactors are eliminated and thus do not contribute to $|\text{cof}_{n,m}(f, I)|$.

If $n$ and $m$ are clear from the context, we simply write $\text{cof}(f, I)$.

Next, we restrict this set to contain only cofactors that are *true* cofactors of one of the single-output functions $f_i$.

DEFINITION 2.8 *Let* $f\colon \mathbf{B}^n \to \mathbf{B}^m$; $f = (f_i^{(n)})_{1\leq i\leq m}$ *be a Boolean multi-output function essentially depending on all its input variables and let* $I \subseteq X_n$. *The set of non-constant true cofactors of* $f$ *with respect to the variables in* $I$ *is denoted* $\mathrm{tcof}_{n,m}(f, I)$. *Formally, a function* $\mathrm{tcof}_{n,m}$ *is given as*

$$\mathrm{tcof}_{n,m}\colon \{f \mid f\colon \mathbf{B}^n \to \mathbf{B}^m\} \times 2^{X_n} \to 2^{\{f \mid f\colon \mathbf{B}^n \to \mathbf{B}^m\}};$$

$$(f, \{x_{i_1}, \ldots, x_{i_k}\}) \mapsto$$
$$\{ f_i|_{x_{i_1}=a_1,\ldots,x_{i_k}=a_k} \textit{ non-constant and true cofactor of } f_i \mid$$
$$1 \leq i \leq m, (a_1, \ldots, a_k) \in \mathbf{B}^k \}$$

*for* $1 \leq k \leq n$.

Let $f_i \neq f_j$ be two distinct single-output functions in the family $(f_i^{(n)})_{1\leq i\leq m}$. Note that a true cofactor in $f_i$ can be functionally equivalent to a cofactor of $f_j$ that is not true. In other words, the cofactors in the set $\mathrm{tcof}_{n,m}(f, I)$ are not required to be true in *every* single-output function, it is only required that at least one such single-output function exists.

Again if $n$ and $m$ are given from the context, we simply write $\mathrm{tcof}(f, I)$.

The following well-known theorem [Sha38] allows to decompose Boolean functions into "simpler" sub-functions.

THEOREM 2.9 *Let* $f\colon \mathbf{B}^n \to \mathbf{B}^m$ *be a Boolean function (over* $X_n$*). For all* $x_i \in X_n$ *we have:*

$$f = x_i \cdot f_{x_i=1} + \overline{x}_i \cdot f_{x_i=0}. \tag{2.1}$$

It is straightforward to see that the sub-functions obtained by subsequent application of Theorem 2.9, called the *Shannon decomposition*, to a function $f$ are uniquely determined. Note that this ensures the well-definedness of cofactor set definitions.

## 2.4 Reduced Ordered Binary Decision Diagrams

Many applications in VLSI CAD make use of reduced ordered *Binary Decision Diagrams* (BDDs) as introduced by [Bry86]:

A BDD is a *graph-based* data structure. Redundant nodes in the graph, i.e. nodes not needed to represent $f$, can be eliminated. BDDs allow a unique (i.e. canonical) representation of Boolean functions. At the same time they allow for a good trade-off between efficiency of manipulation and compactness. Compared to other techniques to represent Boolean functions, e.g. truth tables or Karnaugh maps, BDDs often require much less memory and faster algorithms for their manipulation do exist.

In the following, a formal definition of BDDs is given. We start with purely *syntactical* definitions by means of *Directed Acyclic Graphs* (DAGs). First, single-rooted *Ordered Binary Decision Diagrams* (OBDDs) are defined. This definition is extended to multi-rooted graphs, yielding *Shared* OBDDs (SBDDs). Next, the *semantics* of SBDDs is defined, clarifying how Boolean functions are represented by SBDDs.

After that, reduction operations on SBDDs are introduced which preserve the semantics of an SBDD. This leads to the final definition of reduced SBDDs that will be called BDDs for short in this book.

Finally, some definitions and notations are given which allow to discuss various graph-oriented properties of BDDs, among them paths in BDDs and their (expected) length.

Examples are given to illustrate the formal definitions where appropriate.

### 2.4.1 Syntactical Definition of BDDs

DEFINITION 2.10 *An* Ordered Binary Decision Diagram *(OBDD) is a pair* $(\pi, G)$ *where* $\pi$ *denotes the variable ordering of the OBDD and* $G$ *is a finite DAG* $G = (V, E)$ *(*$V$ *denotes the set of* vertices *and* $E$ *denotes the set of* edges *of the DAG) with exactly one root node (denoted root) and the following properties:*

- *A node in* $V$ *is either a non-terminal node or one of the two terminal nodes in* $\{\mathbf{1}, \mathbf{0}\}$.

- *Each non-terminal node* $v$ *is labeled with a variable in* $X_n$*, denoted* $\mathrm{var}(v)$*, and has exactly two child nodes in* $V$ *which are denoted* $\mathrm{then}(v)$ *and* $\mathrm{else}(v)$.

- *On each path from the root node to a terminal node the variables are encountered at most once and in the same order.*

*More precisely, the variable ordering* $\pi$ *of an OBDD is a bijection*

$$\pi\colon \{1, 2, \ldots, n\} \to X_n$$

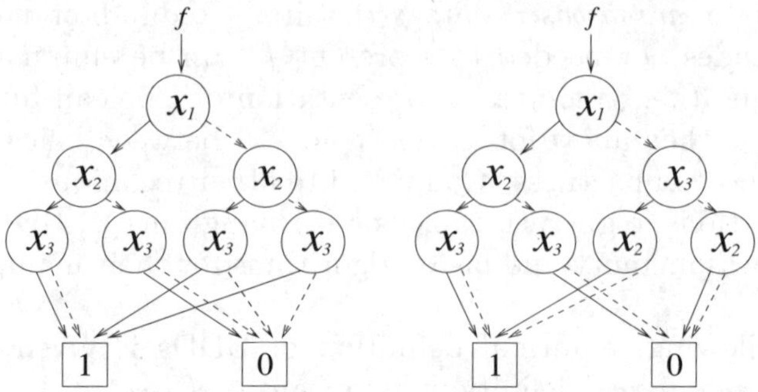


*Figure 2.1.* Representation of $f = x_1 \cdot x_2 + \overline{x}_1 \cdot x_3$ by an OBDD and an unordered BDD.

*where $\pi(i)$ denotes the i-th variable in the ordering. The above condition "in the same order" states that for any non-terminal node $v$ we have*

$$\pi^{-1}(\mathrm{var}(v)) < \pi^{-1}(\mathrm{var}(\mathrm{then}(v)))$$

*iff* then$(v)$ *is also a non-terminal node and*

$$\pi^{-1}(\mathrm{var}(v)) \quad < \quad \pi^{-1}(\mathrm{var}(\mathrm{else}(v)))$$

*iff* else$(v)$ *is also a non-terminal node.*

Even though this might look a bit "over-formal", the notation is required in the following to prove the correctness of the algorithms.

EXAMPLE 2.11 *In Figure 2.1 two different types of binary decision diagrams are depicted. Solid lines are used for the edges from $v$ to* then$(v)$ *whereas dashed lines indicate an edge between $v$ and* else$(v)$. *In both diagrams the variables in $\{x_1, x_2, x_3\}$ are encountered at most once on every path. Whereas the right diagram is not ordered since the variables are differently ordered along distinct paths, the left one respects the ordering $\pi(1) = x_1, \pi(2) = x_2, \pi(3) = x_3$. Both diagrams represent the function $f$: $\mathbf{B}^3 \to \mathbf{B}$; $(x_1, x_2, x_3) \mapsto x_1 \cdot x_2 + \overline{x}_1 \cdot x_3$, as will be explained in a later section.*

Where appropriate, we will speak of an OBDD over $X_n$ to put emphasis on the set of variables used as node labels. However, if not stated otherwise, an OBDD is always assumed to be an OBDD over $X_n$.

Note that OBDDs are *connected* graphs, as all nodes must be connected via at least one path to the (only) root node: to see this, assume the graph consists of more than one connected component. But then, due to the graph being finite and acyclic, there must exist a second root node for the second component graph. This contradicts the assumption that the graph is single-rooted. Let root $\in V$ denote the one root node of the OBDD.

Formally, a *function* var: $V \setminus \{\mathbf{1}, \mathbf{0}\} \rightarrow X_n$ maps non-terminal nodes to variables and *functions* then: $V \setminus \{\mathbf{1}, \mathbf{0}\} \rightarrow V \setminus \{\text{root}\}$ and else: $V \setminus \{\mathbf{1}, \mathbf{0}\} \rightarrow V \setminus \{\text{root}\}$ map non-terminal nodes to child nodes. Both functions are well-defined and total. If not stated otherwise, we will consider surjective functions var, i.e. every variable in $X_n$ appears as label in the OBDD. This corresponds to the convention to consider Boolean functions which essentially depend on every variable in $X_n$. In general none of these functions is required to be injective, i.e. several nodes can have the same label and share the same child nodes. Let $G = (V, E)$ be the underlying graph of an OBDD $(\ldots, G)$ and let $v \in V \setminus \{\mathbf{1}, \mathbf{0}\}$. We call then$(v)$ the 1-child and else$(v)$ the 0-child of $v$.

For an edge $e \in E$ $(E \subseteq V \times V)$, we denote the *type* of the edge with $\text{t}(e)$, i.e. we have $\text{t}(e) = 1$ for an edge $e = (v, \text{then}(v))$ (called a 1-edge) and $\text{t}(e) = 0$ for an edge $e = (v, \text{else}(v))$ (called a 0-edge).

DEFINITION 2.12 *A Shared OBDD (SBDD) is a tuple* $(\pi, G, O)$. $G$ *is a rooted, possibly multi-rooted DAG* $(V, E)$ *which consists of a finite number of graph components. These components are OBDDs, all of them respecting the same variable ordering* $\pi$. $O \subseteq V \setminus \{\mathbf{1}, \mathbf{0}\}$ *is a finite set of output nodes* $O = \{o_1, o_2, \ldots, o_m\}$. *An SBDD has the following properties:*

- *A node in* $V$ *is either a non-terminal node or one of the two terminal nodes in* $\{\mathbf{1}, \mathbf{0}\}$.
- *Every root node of the component OBDD graphs must be contained in* $O$ *(but not necessarily vice versa).*

EXAMPLE 2.13 *An example of an SBDD is given in Figure 2.2. The nodes pointed to by* $f_1, f_2$ *and* $f_3$ *are output nodes. Note that every root node is an output node (pointed to by* $f_2$ *and* $f_3$*). An output node must not necessarily be a root node (see the node pointed to by* $f_1$*).*

Also note that in SBDDs multiple graphs can share the same node, a property which helps to save nodes and to reduce the size of the diagram. In contrast to the OBDD in Figure 2.1, an SBDD represents multiple Boolean functions in only one diagram, as will be explained later.

Where appropriate, again we speak of an SBDD over $X_n$ to clarify the set of node labels.

If there is only one component in $G$ and if $O$ is a singleton (consisting of the one root node of $G$), an SBDD specializes to an OBDD. Later, in

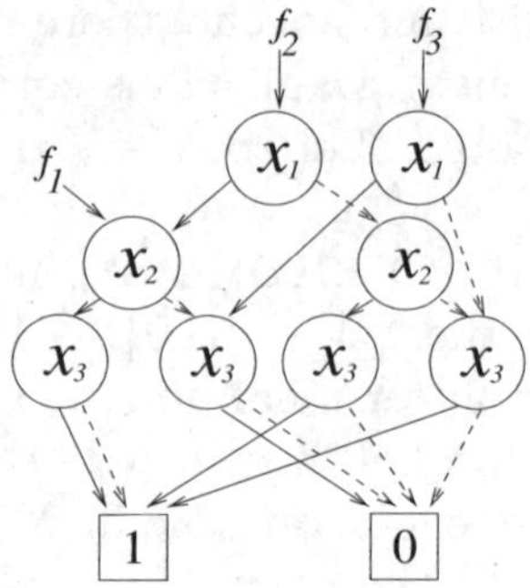


*Figure 2.2.* A shared OBDD (SBDD).

the case that for some reason single-rooted diagrams must explicitly be excluded, or to put emphasis on the multi-rooted case, the term "shared BDD" will be used where a BDD is a "reduced" SBDD (see Definition 2.16 and also Remark 1). Since SBDDs are not necessarily reduced, the terms "SBDD" and "shared BDD" should not be mixed up.

The idea behind the set $O$ is to declare additional non-terminal, non-root nodes as nodes representing Boolean functions. This will be clarified in the next section when the semantics of BDDs is defined. Note that also SBDDs have at most two terminal nodes which are shared by the components.

## 2.4.2 Semantical Definition of BDDs

DEFINITION 2.14 *An SBDD* $(\ldots, G, O)$, *over* $X_n$ *with* $O = \{o_1, o_2, \ldots, o_m\}$ *represents the multi-output function* $f := (f_i^{(n)})_{1 \leq i \leq m}$ *defined as follows:*

- *If* $v$ *is the terminal node* $\mathbf{1}$, *then* $f_v = \text{one}$, *if* $v$ *is the terminal node* $\mathbf{0}$, *then* $f_v = \text{zero}$.
- *If* $v$ *is a non-terminal node and* $\text{var}(v) = x_i$, *then* $f_v$ *is the function*

$$\begin{aligned} & f_v(x_1, x_2, \ldots, x_n) \\ = \quad & x_i \cdot f_{\text{then}(v)}(x_1, x_2, \ldots, x_n) + \overline{x}_i \cdot f_{\text{else}(v)}(x_1, x_2, \ldots, x_n). \end{aligned}$$

- *For* $1 \leq i \leq m$, $f_i$ *is the function represented by the node* $o_i$.

The expression $f_{\text{then}(v)}$ $(f_{\text{else}(v)})$ denotes the function represented by the child nodes then$(v)$ (else$(v)$). At each node of the SBDD, essentially a Shannon decomposition (see Theorem 2.9) is performed. In this, an SBDD recursively splits a function into simpler sub-functions. The first

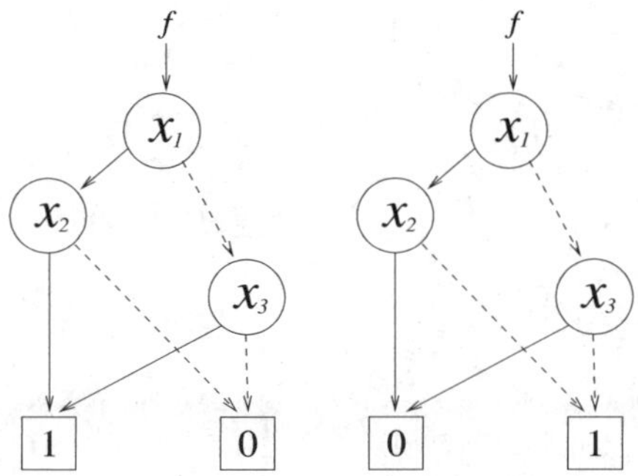


*Figure 2.3.* Representation of $f = x_1 \cdot x_2 + \overline{x}_1 \cdot x_3$ and $\overline{f} = x_1 \cdot \overline{x}_2 + \overline{x}_1 \cdot \overline{x}_3$.

definitions of binary decision diagrams are due to [Lee59] and [Ake78a]. *Complemented Edges* (CEs) are an important extension of the basic BDD concept and have been described in [Ake78b, Kar88, MB88, MIY90, BRB90]. The idea is to use the close similarity between the BDD representing a function $f$ and the BDD representing its complement. For example see Figure 2.3 where a left BDD for $f = x_1 \cdot x_2 + \overline{x}_1 \cdot x_3$ and a right BDD for $\overline{f} = x_1 \cdot \overline{x}_2 + \overline{x}_1 \cdot \overline{x}_3$ is given. The diagrams are identical except for interchanged terminal nodes.

A CE is an ordinary edge that is tagged with an extra bit (*complement bit*): This bit is set to indicate that the connected sub-graph must be interpreted as the complement of the formula that the sub-graph represents. CEs allow to represent both a function and its complement by the same node, modifying the edge pointing to that node instead. As a consequence, only one constant node is needed. Usually the node **1** is kept, allowing the function zero to be represented by a CE to **1**.

Note that it is not necessary to store the complement bit as an extra element of the node structure. Instead, a smart implementation technique can be used which exploits the memory alignment used in present computer systems: modern CPUs require allocated objects to reside at a memory address that is evenly divisible by some small constant, e.g. often this is the constant two or four. Consequently, the least significant bit of a pointer word is irrelevant for address calculation. Therefore it can be used to store the complement bit (provided that an appropriate bit mask is applied before address calculation). Hence using CEs does not cause any memory overhead. SBDDs with CEs are used today in many modern BDD packages.

For the sake of a more understandable presentation of the achieved results, in this book "classical", unmodified SBDDs without CEs are used in most of the examples and their illustrations (with the exception

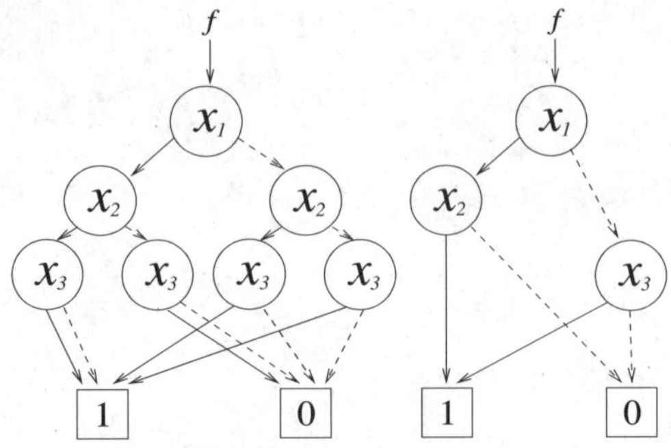


*Figure 2.4.* Two different SBDDs for $f$: $(x_1, x_2, x_3) \mapsto x_1 \cdot x_2 + \overline{x}_1 \cdot x_3$.

of Section 5.1 where CEs are discussed explicitly). However, all results presented throughout the book hold for BDDs with CEs and can easily be transferred to BDDs without CEs.

Note that in case of using CEs special attention must be drawn to maintain a so-called "canonical form", i.e. to achieve uniqueness of the representation of Boolean functions. This will be addressed again in Section 2.4.3 where the unique, irreducible form of BDDs is introduced.

### 2.4.3 Reduction Operations on SBDDs

In the previous sections, we developed syntax and semantics of a special form of graphs which are representing Boolean functions. However, even if the considered variable ordering is fixed, still there exist several possibilities of representing a given function: in Figure 2.4 we see two different SBDDs respecting the same variable ordering $\pi^{-1}(x_1) < \pi^{-1}(x_2) < \pi^{-1}(x_3)$, representing the same function $f$: $\mathbf{B}^3 \to \mathbf{B}$; $(x_1, x_2, x_3) \mapsto x_1 \cdot x_2 + \overline{x}_1 \cdot x_3$. Next, we will give reduction operations on SBDDs which transform an SBDD into an irreducible form, while the function represented by the SBDD is preserved. This is a crucial technique leading to an SBDD respecting a given variable ordering which is unique up to graph isomorphism and of minimal size.

A *reduced* SBDD then is the final form of binary decision diagrams that will be considered throughout this book, called BDD.

DEFINITION 2.15 *Given an SBDD* $(\ldots, G, O)$, *there are the following two* reduction rules *which can be applied to* $G = (V, E)$:

- **Deletion Rule:** *Let* $v \in V$ *be a node with*

$$\text{then}(v) = \text{else}(v) =: v'.$$

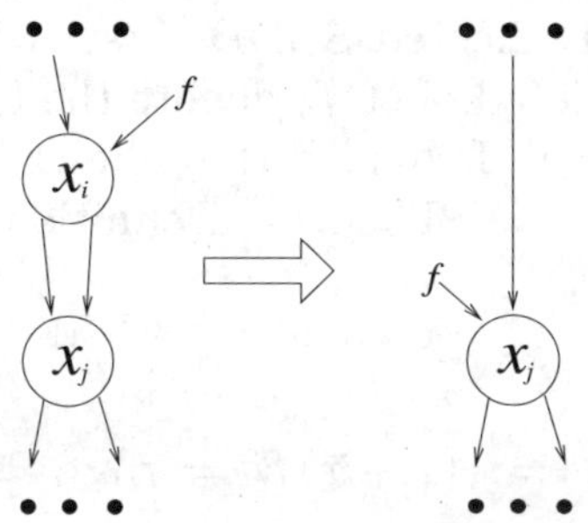


*Figure 2.5.* Deletion Rule for SBDD-reduction.

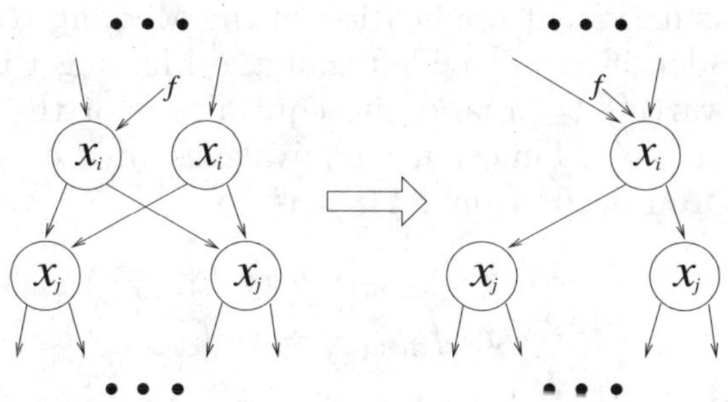


*Figure 2.6.* Merging Rule for SBDD-reduction.

*Then the reduction* $\text{del}(v, v')$ *is applicable to* $G$*, resulting in a graph* $G[\text{del}(v, v')]$ *obtained by a) redirecting all edges that are pointing to* $v$ *to the node* $v'$ *and b) inserting* $v'$ *into* $O$ *iff* $v \in O$ *and c) deleting* $v$ *from* $V$ *and* $O$*.*

- **Merging Rule:** *Let* $v, v' \in V$ *be two nodes with*

  *1)* $\text{var}(v) = \text{var}(v')$,

  *2)* $\text{then}(v) = \text{then}(v')$, *and*

  *3)* $\text{else}(v) = \text{else}(v')$.

  *Then the reduction* $\text{merge}(v, v')$ *is applicable to* $G$*, resulting in a graph* $G[\text{merge}(v, v')]$ *obtained by a) redirecting all edges pointing to* $v$ *to the node* $v'$ *and b) inserting* $v'$ *into* $O$ *iff* $v \in O$ *and c) deleting* $v$ *from* $V$ *and* $O$*.*

In both rules with action b) we ensure that output nodes do not "vanish". In Figures 2.5 and 2.6 the application of both reduction rules is illustrated. It is straightforward to see that the application of one of the two reduction operations del and merge does not change the function represented by the affected SBDD: first, with the Deletion Rule, redundant nodes are deleted. Let $f_v$ denote the function represented by node $v$ and let $\text{var}(v) = x$. Provided that the Deletion Rule applies, we have $\text{then}(v) = \text{else}(v) =: v'$. Then, by Definition 2.14,

$$\begin{aligned} f_v &= x \cdot f_{\text{then}(v)} + \overline{x} \cdot f_{\text{else}(v)} \\ &= x \cdot f_{v'} + \overline{x} \cdot f_{v'} \\ &= (x + \overline{x}) f_{v'} = f_{v'}, \end{aligned}$$

hence deletion of $v$ and redirection of all ingoing edges to the functionally equivalent node $v'$ preserves the function represented by the diagram.

Second, with subsequent application of the Merging Rule isomorphic sub-graphs are identified. Provided that the Merging Rule applies, we have $\text{var}(v) = \text{var}(v') := x$ and the equivalence of the 1-children (0-children) of $v$ and $v'$. Functional equivalence of the nodes $v$ and $v'$ follows directly from Definition 2.14: it is

$$\begin{aligned} f_v &= x \cdot f_{\text{then}(v)} + \overline{x} \cdot f_{\text{else}(v)} \\ &= x \cdot f_{\text{then}(v')} + \overline{x} \cdot f_{\text{else}(v')} \\ &= f_{v'}, \end{aligned}$$

thus using $v'$ instead of $v$ on every path of the BDD and discarding $v$ collapses the graph but the represented function is not changed.

DEFINITION 2.16 *An SBDD is called* reduced *iff there is no node where one of the reduction rules (i.e., the Deletion or the Merging Rule) applies.*

The term "reduced" is used here since it has become common in the literature. This is done despite the fact that the older and more accurate term "irreducible" known from the paradigm of rewriting systems could be applied as well.

REMARK 1 *In the following, only reduced SBDDs are considered and for simplicity, they are called BDDs (or shared BDDs when explicitly excluding single-rooted BDDs).*

The *size of a BDD* $F = (\ldots, G, \ldots)$ is the number of nodes in the underlying graph $G$, denoted $|F|$ (or sometimes also $|G|$). The following theorem [Bry86] holds:

THEOREM 2.17 *BDDs are a canonical representation of Boolean functions, i.e. the BDD-representation of a given Boolean function with respect to a fixed variable ordering is unique up to isomorphism and is of minimal size.*

This property of BDDs is useful especially when checking the equivalence of two Boolean functions represented as a BDD. If the functions are indeed equivalent, their representation must be isomorphic. Actually, in modern, efficient BDD packages [BRB90, DB98, Som01, Som02] equivalent functions are represented by the same graph. Once the BDDs for the functions have been built, equivalence checking can be done in constant time (comparing the address of the root nodes of the representing graphs).

In a similar manner, it can be decided in constant time whether a Boolean function is satisfiable or not, once the BDD representing the function has been built: it suffices to check whether the BDD is the one for the constant zero function or not. In the latter case, the function must be satisfiable.

Moreover, it is known that computing a satisfying assignment of the $n$ input variables can be done in time which is linear in $n$ [Bry86].

The result of Theorem 2.17 is reflected in the next definition introducing a convenient notation to refer to the *one* BDD representing a Boolean function $f$ with respect to a variable ordering $\pi$: we use the fact that the set of all BDDs representing a given Boolean function $f^{(n,m)}$ can be decomposed into equivalence classes of isomorphic BDDs, i.e. we have one class for every ordering.

DEFINITION 2.18 *Let* $f^{(n,m)}$ *be a Boolean function and let* $\pi^{(n)}$ *a variable ordering. Then* $\mathrm{BDD}_{n,m}(f,\pi)$ *denotes the equivalence class of BDDs representing* $f$ *and respecting* $\pi$. *If* $n$, $m$ *are clear, we simply write* $\mathrm{BDD}(f,\pi)$.

We may find it convenient to *identify* a class $\mathrm{BDD}(f,\pi)$ with an arbitrary *representative*, i.e. we speak of *the* BDD $\mathrm{BDD}(f,\pi)$ instead of the class of all BDDs isomorphic to the (chosen) representative. Formally, if $\overline{F}$ is a class of BDDs, we identify $\overline{F}$ with $F$.

An example of this would be using $\mathrm{BDD}(f,\pi)$ as parameter of a function which is expecting a BDD as argument. This simplification is harmless, as long as the following holds: the resulting function value must be preserved with respect to changes of the class representative. Instead of declaring a BDD with $F \in \mathrm{BDD}(f,\pi)$, this will be done in the form $F := \mathrm{BDD}(f,\pi)$. This is a relaxed notation, expressing that $F$ is assigned a certain, fixed class representative. This notation is used to support complete abstraction from the details of graph isomorphism.

The introduced terminology directly transfers to BDDs with CEs. However, as has already been mentioned before in Section 2.4.2, special care has to be taken here to maintain a canonical form. The reason is the existence of several functional equivalences of distinct, reduced

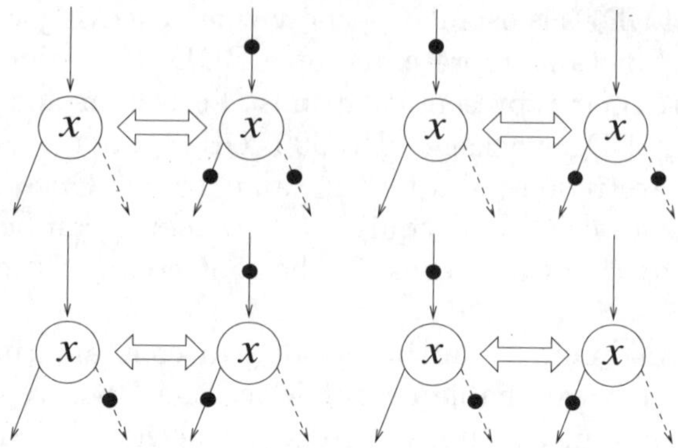


*Figure 2.7.* Pairs of functions represented as BDDs with CEs that are functional equivalent.

BDDs with CEs as illustrated in Figure 2.7. A dot on an edge indicates a complemented edge (otherwise it is a regular edge).

As a remedy, it must be constrained where CEs are used. To achieve a unique representation it suffices to restrict the 1-edge of every node to be a regular edge. Thus, in Figure 2.7 always the left member of each functionally equivalent pair is chosen. It can be shown that this already guarantees a canonical form. All function-preserving reduction operations which follow this constraint result in a unique BDD with CEs which also respects this condition.

### 2.4.4 Variable Orderings of BDDs

A problem with BDDs however is their *sensitivity to variable ordering*. To illustrate this, an example is reviewed which has been given in [Bry86].

EXAMPLE 2.19 *Let $n \in \mathbb{N}$ be even and let $f$: $\mathbf{B}^n \rightarrow \mathbf{B}$;*

$$(x_1, x_2, \ldots, x_n) \mapsto x_1 \cdot x_2 + x_3 \cdot x_4 + \ldots + x_{n-1} \cdot x_n.$$

*A BDD for $f$ respecting the variable ordering $\pi(1) = x_1, \pi(2) = x_2, \ldots, \pi(n-1) = x_{n-1}, \pi(n) = x_n$ is given in Figure 2.8: the left BDD is of size $n$. Since $f$ obviously depends on all $n$ variables, this is the optimal size.*

*However, the right BDD for $f$ respecting the variable ordering $\pi(1) = x_1, \pi(2) = x_3, \ldots, \pi(n/2) = x_{n-1}, \pi(n/2+1) = x_2, \pi(n/2+2) = x_4, \ldots, \pi(n) = x_n$ is of exponential size: it is straightforward to see that the graph is of size $2^{(n/2)+1} - 2$.*

That is, depending on the variable ordering the size of a BDD may vary from linear to exponential in $n$, the number of input variables.

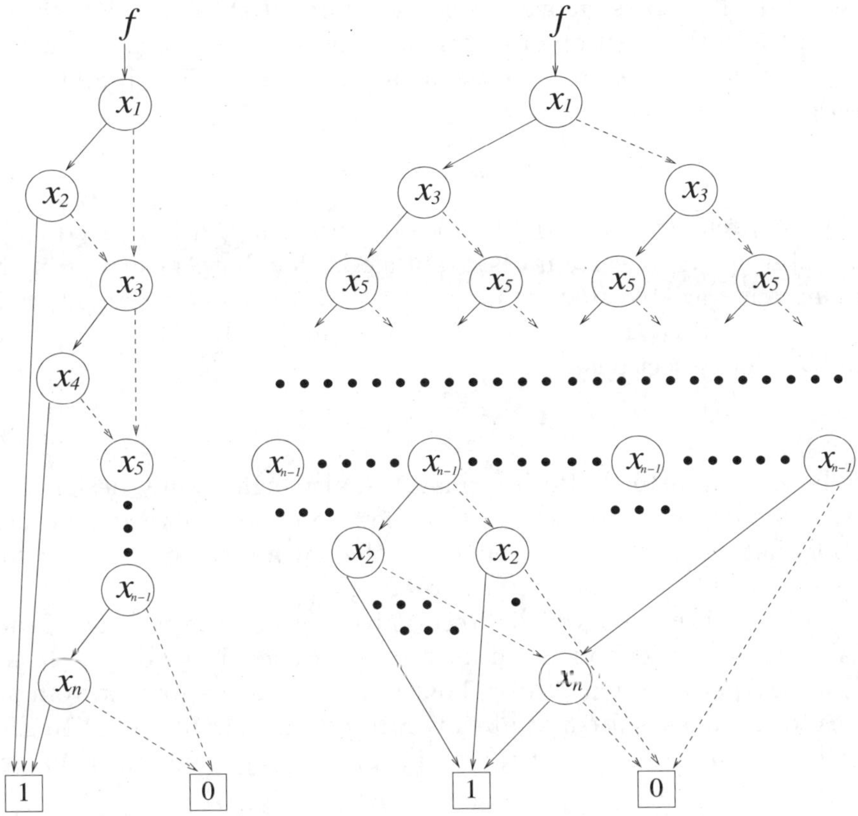


*Figure 2.8.* Two BDDs for $f = x_1 \cdot x_2 + x_3 \cdot x_4 + \ldots + x_{n-1} \cdot x_n$.

To describe approaches to minimization of BDD size (as will be done in Chapters 3 and 4), a formalism expressing variable orderings, changes of these orderings and movements of variables is necessary. This section introduces the notation used throughout this book for this purpose.

In some cases, when considering the variable ordering of a given BDD, the ordering is not needed explicitly. E.g. this is the case, when a BDD is given with respect to an initial ordering: considering methods for BDD size minimization, we are often interested in expressing (or restricting) the effect of the changes from one ordering to another. In this case we are not interested in the initial ordering from which a heuristic or an exact reordering method has been started.

In such cases we omit stating the ordering and thus in all examples, if not stated otherwise, the "natural" ordering defined by

$$\pi(i) = x_i \quad (i \in \{1, 2, \ldots, n\})$$

is assumed. To express movements (shifts) of variables in the ordering, i.e. changes in the position of a variable and exchanging a variable with another variable, often *permutations* have been used. E.g. [FS90], used bijections

$$\pi\colon \{1, 2, \ldots, n\} \to \{1, 2, \ldots, n\}$$

to express exchanges of variables. Here, variables $x_i$ are referred to by their subscript $i$. A certain disadvantage is that both positions in the ordering and variable subscripts are denoted using natural numbers in $\{1, 2, \ldots, n\}$, hence they can be mixed up quite easily.

In [DDG00] bijections

$$\pi\colon X_n \to X_n$$

have been used instead. Both formalisms, although mainly designed to express *changes* in an ordering, also allow to express a variable ordering: for that we assume $\pi$ is applied to the natural ordering, the result is then assumed to be the current ordering.

However, a formalism which directly gives the current position of each variable can be easier to read in most cases. Hence, instead of denoting orderings as permutations, throughout the book the mapping $\pi$ from positions to variables is used, as already introduced in Definition 2.10. For the outlined historical reasons (see above), this bijection still is named $\pi$.

The next definition will be used later in Chapter 3 to express sets of orderings which respect a certain condition.

DEFINITION 2.20 *Let $I \subseteq X_n$. Then $\Pi_n(I)$ denotes the set*

$$\Pi_n(I) = \{\pi\colon \{1, 2, \ldots, n\} \to X_n \mid \{\pi(1), \pi(2), \ldots, \pi(|I|)\} = I\}.$$

*A mapping $\pi \in \Pi_n(I)$ is a variable ordering whose first $|I|$ positions constitute $I$. Normally we omit the arity $n$, writing $\Pi(I)$ instead of $\Pi_n(I)$, if $n$ is given from the context.*

This allows us to focus the attention to variable orderings with the following property: an ordering $\pi^{(n)} \in \Pi_n(I)$ partitions the set of variables $X_n$ into a partition $(L, R)$: let $(\pi, G, \ldots)$ be a BDD over $X_n$. Nodes with a label in $L = I$ reside in the upper part of $G$, i.e. in levels $1, 2, \ldots, |I|$. Nodes with a label in $R = X_n \setminus I$ reside in the lower part of $G$ in levels $|I| + 1, |I| + 2, \ldots, n$. The definition also allows us to force the last $|I|$ positions of an ordering to constitute $I$: this holds for all $\pi \in \Pi(X_n \setminus I)$.

Here, the term "level" has been used informally. In the next section a formal definition of the term "BDD level" will be given.

### 2.4.5 BDD Levels and their Sizes

In this section additional notations are given which are needed to formally refer to parts of the BDD, to their sizes and to minimal sizes.

DEFINITION 2.21 *A* level *of a BDD over $X_n$ (or, equivalently, a BDD level) is a set containing all non-terminal nodes labeled with one particular variable $x_i$ in $X_n$. Let $F = (\pi^{(n)}, \ldots)$ be a BDD over $X_n$. If $x_i = \pi(k)$, we speak of the "$x_i$-level" as well as the "$k$-th level" or "level $k$".*

DEFINITION 2.22 *Let*

$$\text{nodes}_n\colon \{F \mid F \textit{ is a BDD over } X_n\} \times X_n \to 2^{\{v \in V \mid (\ldots,(V,E),\ldots) \textit{ is a BDD}\}};$$

$$\text{nodes}_n(F, x_i) = \{v \mid v \in V, \text{var}(v) = x_i \textit{ where } F = (\ldots, (V, E), \ldots)\}.$$

We straightforwardly extend this definition to also cover sets of variables, i.e. we define

$$\text{nodes}_n\colon \{F \mid F \text{ is a BDD over } X_n\} \times 2^{X_n} \to 2^{\{v \in V \mid (\ldots,(V,E),\ldots) \text{ is a BDD}\}};$$

$$\text{nodes}_n(F, X) = \bigcup_{x_i \in X} \text{nodes}_n(F, x_i).$$

For simplicity, this extension is denoted with the same function symbol. If $n$ is clear from the context, we omit the subscript $n$, writing nodes$(F, x_i)$ and nodes$(F, X)$ instead of nodes$_n(F, x_i)$ and nodes$_n(F, X)$. The term nodes$(F, x_i)$ denotes the set of nodes in the $x_i$-level of $F$ whereas the term nodes$(F, X)$ denotes the set of nodes in $F$ labeled with a variable in $X \subseteq X_n$.

To express sizes of levels, i.e. the number of nodes in a level, we also introduce the following function.

DEFINITION 2.23 label$_n$: $\{F \mid F \textit{ is a BDD over } X_n\} \times X_n \to \mathbb{N}$;

$$\text{label}_n(F, x_i) = |\text{nodes}_n(F, x_i)|\,,$$

*and* label$_n$: $\{F \mid F \textit{ is a BDD over } X_n\} \times 2^{X_n} \to \mathbb{N}$;

$$\text{label}_n(F, X) = |\text{nodes}_n(F, X)|\,.$$

As before, the subscript $n$ is ommited if $n$ is given by the context and the same function symbol is used both for the basic and the extended function. Note that the size of the $k$-th level can be expressed as label$(F, \pi(k))$. To be able to refer to the set of nodes in the $k$-th level in a BDD, we introduce the following definition.

DEFINITION 2.24

$$\text{level: } \{F \mid F \textit{ is a BDD}\} \times \mathbb{N} \to 2^{\{v \in V \mid (\ldots,(V,E),\ldots) \textit{ is a BDD}\}};$$

$$\text{level}(F, k) = \{v \mid v \in V, \text{var}(v) = \pi(k) \textit{ where } F = (\pi, (V, E), \ldots)\}.$$

Let $F = (\ldots, \ldots, O)$ be a BDD. To express the set of nodes in parts of $F$ covering (the output nodes situated at) several levels we introduce the following notations. Let

$$F_j^i = \bigcup_{i \leq k \leq j} \text{level}(F, k)$$

and let

$$O_j^i = O \cap F_j^i.$$

Due to the ordering restriction imposed on the variables of a BDD, it is possible to levelize each BDD graph illustration, i.e. to rectify the graph such that all nodes with the same label appear at the same level of height in the graph. Later, in Chapter 3, we will need to express the minimal number of nodes in BDDs or parts of BDDs. This can be done using the following very flexible definition.

DEFINITION 2.25 *Let $f$ be an $n$-ary Boolean function and let $I, J \subseteq X_n$.*

$$\text{min\_cost}_f(I, J) = \min_{\pi \in \Pi(J)} (\text{label}(\text{BDD}(f, \pi), I)),$$

*i.e., $\text{min\_cost}_f(I, J)$ denotes the minimal number of nodes labeled with a variable in $I$ under all BDDs representing $f$ with a variable ordering whose first $|J|$ elements constitute $J$. If the function $f$ is given from the context, we omit it, writing* $\text{min\_cost}(I, J)$ *for short.*

If $J = I$, all orderings are considered which change the order of variables within the upper part of the BDD in levels $1, 2, \ldots, |I|$ such that all nodes in this part remain labeled with a variable in $I$.

The term $\text{min\_cost}(I, I)$ expresses the size of the upper part for a "best" of all these orderings, i.e. the minimal size of the upper part with respect to a partition $(I, X_n \setminus I)$ of the set $X_n$.

If $J = X_n \setminus I$, all orderings are considered which change the order of variables within the *lower* part of the BDD in levels $n - |I| + 1, n - |I| +$

$2, \ldots, n$ such that all nodes in this part remain labeled with a variable in $I$.

The term $\text{min\_cost}(I, X_n \setminus I)$ expresses the size of the *lower* part for a "best" of all these orderings, i.e. the minimal size of the lower part with respect to a partition $(X_n \setminus I, I)$ of the set $X_n$.

Instead, we can also consider a partition $(I, X_n \setminus I)$, again considering all orderings which change the order of variables within the upper part of the BDD in levels $1, 2, \ldots, |I|$ and then express the minimal size of the *lower part* of the BDD under all these orderings: this would be expressed by the term $\text{min\_cost}(X_n \setminus I, I)$.

Yet, besides these possibilities, the definition allows to express even more sophisticated minimal sizes of BDD parts. But throughout this book, only the above forms will be needed and used.

Formally, $\text{min\_cost}_f$ is a function

$$\text{min\_cost}_f \colon 2^{X_n} \times 2^{X_n} \to \mathbb{N}.$$

The well-definedness follows from Theorem 2.17 which ensures that, for $I \subseteq X_n$, the term $\text{label}(\text{BDD}(f, \pi), I)$ is uniquely determined by $f$ and $\pi$.

### 2.4.6 Implementation of BDD Packages

In the previous sections, the concept of BDDs has been introduced. In practice, the success of BDD-based algorithms relies on the efficient implementation of this concept. This section describes the basic implementation techniques used today by modern BDD packages. For more details see [BRB90, Som01].

In Section 2.4.3 it was explained why for typical applications like a functional equivalence check only reduced BDDs are of interest. For this reason it is desirable to avoid generating unreduced BDDs during the operation of a BDD package. A way of achieving this is to check whether a node representing a particular function already exists. This is done before the creation of a node representing this function. For a node $v$, the function represented by $v$, $f_v$, is uniquely determined by the tuple

$$(\text{var}(v), \text{then}(v), \text{else}(v)),$$

containing as its elements all arguments of the Shannon decomposition (see Theorem 2.9) carried out at $v$:

$$f_v = x \cdot f_{\text{then(v)}} + \overline{x} \cdot f_{\text{else(v)}}$$

where $x = \text{var}(v)$. Of course it would be too time-consuming to compare the tuple of a new node to all tuples of already existing nodes. Hence,

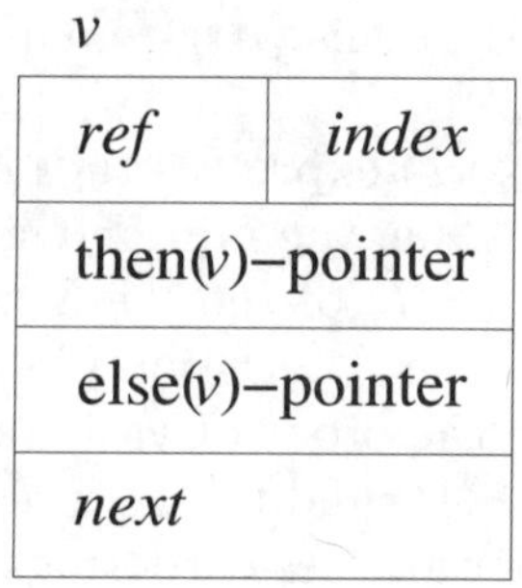


*Figure 2.9.* Node structure.

a global hash table, called the *unique table* is used which allows to find the node $v$ for a tuple $(x, v_1, v_0)$ in constant time. A hash function is computed on the tuple returning an index into an array of bins each storing the first node for that hash value. All other nodes with the same hash value are stored in a collision list. Usually, a next-pointer in the node structure is used to implement this linked list. Other pointers stored at each node are the pointers to the 1-child and the 0-child. Note that only forward-pointers are used to form the DAG, as backward-pointers would increase the node structure and cause too much memory overhead. Besides this, also the variable index *index* (plus flags, i.e. the complement bit, see Section 2.4.3) and a reference count *ref* are stored in the node structure (see Figure 2.9).

The reference count *ref* gives the information how often a node is used at the moment. That way it is possible to free a node if it is not used anymore. This allows the efficient usage of the memory. The count *ref* is updated dynamically whenever the number of references from other nodes or returned user functions changes. If *ref* has reached its maximum value, it is frozen, i.e. the node is not deleted during program run and exists until program termination.

If *ref* reaches the value zero for a node $v$, the node is called a *dead node.* The reference counts of the descendants of $v$ need to be decreased recursively. However, the memory of the dead node $v$ is not freed immediately, because there might still exist an entry in the so-called *computed table* which points at $v$.

The computed table is a global cache storing the intermediate and final results of the recursive algorithms which operate on the DAG of the BDD. A result in this context means a node of the DAG which is the root of a sub-graph representing the computed function.

The idea is to trade memory vs. run time: if a result can be found in the cache, it has been computed before and it is avoided to compute it again. As the number of results can be high during a program run with many BDD operations, the computed table would soon grow too large if implemented as a standard hash table. Therefore usually it is implemented as a *hash-based cache.*

Though the result of a new computation is stored in the cache (again, in form of a tuple pointing to a node via hashing), the handling is different from that of the unique table: rather than appending a new entry to a collision chain, an entry is simply overwritten if a collision occurs. That way a good trade-off between memory requirement and run time improvement can be achieved. Note that it is no problem if a result node is a dead node. In this case a *reclaim* operation, increasing the reference count, re-establishes the node. The reclaimed node is not a dead node anymore.

The reference count also allows for an efficient memory management of the unique table: if a new node is created causing the unique table to become too full, then there are two possible ways of remediation: as a first possibility a *garbage collection* can be performed. This is only done if the number of dead nodes exceeds a certain threshold. Otherwise freeing the dead nodes would not result in a sufficient amount of memory to be recycled, not making the garbage collection worthwile. During garbage collection, dead nodes are freed and entries in the computed table pointing to dead nodes are deleted. Second, if not enough dead nodes are available, both the unique table array and the computed table cache are increased in size, typically by a factor of two. As a consequence, all entries must be assigned a new valid position. This happens using the hash function whose return values have now changed due to the increase of the table size. This process is called *rehashing.* During rehashing, typically also an implicit garbage collection is performed.

Since the tables are increased by a factor of two every time, the rehashing operation is involved only logarithmically often. If both methods to obtain more memory, garbage collection and table size increase with rehashing, fail, the system returns a NULL pointer to the user. At this point it is up to the user to save space by freeing nodes.

Another important implementation aspect used by modern BDD packages is the use of an array of unique tables, one per level $i = 1, \dots n$. Index $i$ serves to locate the hash table which stores all nodes for level $i$. Then all nodes of level $i$ can be traversed by stepping through all collision chains, each starting at a hash table array position.

### 2.4.7 Basic Minimization Algorithm

Algorithms for the minimization of BDDs make use of basic techniques such as *variable swaps* and the *sifting* algorithm. This method has been described in [Rud93] and exploits the only local complexity of the swap operation [ISY91] to achieve good run times. This section gives both basic techniques as the algorithms for BDD optimization presented in Chapters 3, 4, and 5 are based on them.

#### 2.4.7.1 Variable swap

This section describes how a swap of adjacent variables can be implemented such that it is performed with a run time which is proportional to the sizes of the neighboring levels at which the variables are situated, i.e. a *local* behavior.

A problem which is encountered when trying to achieve a local behavior during a swap is the following: assume that a variable swap results in the change of the function represented by a BDD node $v$. As a consequence, all previous references to $v$ in the DAG are not valid anymore and need to be corrected. In Section 2.4.6 it has already been described that modern BDD packages do not maintain back-pointers (from child nodes to parent nodes) in order to reduce memory requirement. Hence, to find all references to $v$ from nodes situated above, we would have to traverse large parts of the DAG, and even then the problem to patch the references from user functions still remains.

For this reason it is necessary to preserve the function represented at each node involved in a swap. This is achieved as follows: instead of constructing a new node in another part of the computer's memory, the data of the node in question (i.e. its tuple, see Section 2.4.6) is overwritten with new values in a function preserving manner. These new values result from the change in the variable ordering. Besides avoiding the problem described above, another advantage of this schema is that no time-consuming memory allocations are needed.

In detail, this is done as follows. Assume a natural variable ordering and assume that variables $x_i = \pi(i)$ and $x_{i+1} = \pi(i+1)$ are swapped. Let $v$ be a node situated at level $i$ and let $v_1$, ($v_0$) denote the 1-child (0-child) of $v$.

Let $f_v$ denote the function represented by $v$ and let $f_1$ ($f_0$) denote the positive (negative) cofactor of $f_v$ with respect to $x_i$. Clearly, $f_1$ ($f_0$) is represented by $v_1$ ($v_0$). Further, let $f_{11}$, ($f_{10}$) denote the positive (negative) cofactor of $f_1$ with respect to $x_{i+1}$. Similarly, let $f_{01}$, ($f_{00}$) denote the positive (negative) cofactor of $f_0$ with respect to $x_{i+1}$. Let $v_{11}$ be the node representing $f_{11}$. This is either the 1-child of $v_1$ (if

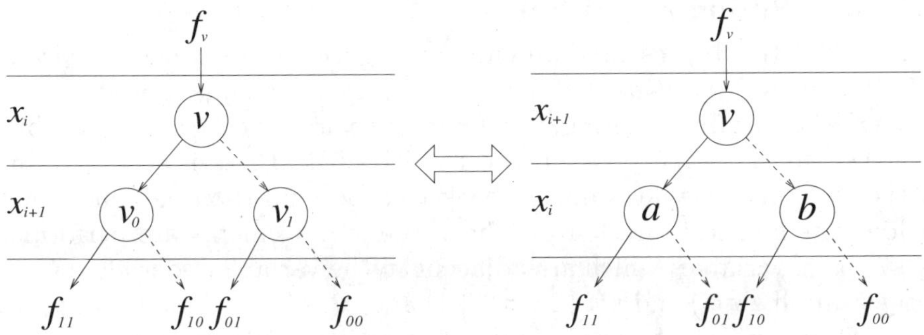


*Figure 2.10.* Variable swap.

$x_{i+1}$ is tested at $v_1$) or simply $v_1$, otherwise. Similarly, let $v_{10}$, and $v_{01}$, $v_{00}$ be the nodes representing $f_{10}$, and $f_{01}$, $f_{00}$, respectively. The left side of Figure 2.10 illustrates the most general case where all cofactors are distinct functions. If some of the cofactors $f_{11}$, $f_{10}$, $f_{01}$, and $f_{00}$ are functionally equivalent, the respective representing nodes have been collapsed by a merge operation before and thus less nodes are involved in the swap than depicted in Figure 2.10.

Node $v$ is represented by the tuple $(x_i, v_1, v_0)$. Performing the swap of $x_i$ and $x_{i+1}$, $v$ is overwritten by $(x_{i+1}, (x_i, v_{11}, v_{01}), (x_i, v_{10}, v_{00}))$. Inspecting the paths through $v$ shows that the new variable ordering (i.e., $x_{i+1}$ "above" $x_i$) is established (see the right side of Figure 2.10). Moreover, this modification preserves the function represented by $v$:

$$
\begin{aligned}
&(x_{i+1}, (x_i, v_{11}, v_{01}), (x_i, v_{10}, v_{00}))\\
&\quad= (x_{i+1}, x_i \cdot f_{11} + \overline{x}_i \cdot f_{01}, x_i \cdot f_{10} + \overline{x}_i \cdot f_{00})\\
&\quad= x_{i+1} \cdot (x_i \cdot f_{11} + \overline{x}_i \cdot f_{01}) + \overline{x}_{i+1} \cdot (x_i \cdot f_{10} + \overline{x}_i \cdot f_{00})\\
&\quad= x_i \cdot x_{i+1} \cdot f_{11} + \overline{x}_i \cdot x_{i+1} \cdot f_{01} + x_i \cdot \overline{x}_{i+1} \cdot f_{10} + \overline{x}_i \cdot \overline{x}_{i+1} \cdot f_{00}\\
&\quad= x_i \cdot (x_{i+1} \cdot f_{11} + \overline{x}_{i+1} \cdot f_{10}) + \overline{x}_i \cdot (x_{i+1} \cdot f_{01} + \overline{x}_{i+1} \cdot f_{00})\\
&\quad= x_i \cdot f_1 + \overline{x}_i \cdot f_0\\
&\quad= f_v
\end{aligned}
$$

The situation after the swap is illustrated on the right side of Figure 2.10. Nodes $v_0$ and $v_1$ will only vanish, if no other (external or user) reference besides those from node $v$ existed. Nodes $a$ and $b$ must be newly created only if nodes with their tuples did not already exist, otherwise they are simply retrieved from the unique table (see Section 2.4.6).

#### 2.4.7.2 Sifting Algorithm

In 1993, Rudell presented an effective algorithm to reduce the size of a BDD, the *sifting algorithm* [Rud93]. Every variable is moved up and down in the variable ordering, i.e. the relative order of the other variables is preserved. At each position the resulting BDD size is recorded and in the end the variable is moved back to a best position, i.e. one which yielded the smallest BDD size. The variable movements are performed by swaps of variables which are adjacent in the variable ordering, e.g. for $i < j$ swapping $\pi(i)$ and $\pi(i+1)$, $\pi(i+1)$ and $\pi(i+2)$ ..., finally swapping $\pi(j-1)$ and $\pi(j)$ moves $\pi(i)$ to the $j$-th position in the ordering. Note that $\pi$ changes with each swap.

In Section 2.4.7.1 it has been shown that a swap of adjacent variables only affects the graph structure of the two levels involved in the swap. Since the number of nodes which have to be touched directly transfers to the run time of the method, this *locality* of the swap operation is a main reason for the efficiency of the sifting algorithm: these exchanges are performed very quickly since only edges must be redirected within these two levels. Moreover, changes in the graph structure can often be established simply by updates of the node data, thus saving the cost of many memory allocations. In this, sifting is based on effective local operations. Moreover, further reducing the number of swaps obviously yields reductions in the run time of the method.

In the past, several methods of achieving reductions in the number of variable swaps have been suggested. A trivial method for example is to start with moving the variable to the *closest end first* when moving a variable to all other positions in the relative order of the variables.

A much more sophisticated method is the use of lower bounds during sifting, which will be discussed in more detail in Chapter 4.

Another improvement suggested is the idea to start with the variable situated at the level with the largest number of nodes: this helps to reduce the overall BDD size as early as possible, thus reducing the complexity of all subsequent steps.

Summarized, the classical sifting algorithm works as follows:

1 The levels are sorted according to their sizes. The largest level is considered first.

2 For each variable:

   (a) The variable is first moved downwards if it is situated closer to the bottommost variable, otherwise it is first moved upwards. Moving the variable means exchanging it repeatedly with its successor variable if moved downwards or with its predecessor variable if moved upwards. This process stops, when the variable has become the bottommost variable (if moved downwards) or the topmost variable (if moved upwards).

(b) The variable is moved into the opposite direction of the previous step, i.e. if in the previous step the variable has been moved downwards, now it is moved upwards until it has become the topmost variable. If, however, in the previous step the variable has been moved upwards, now it is moved downwards until it has become the bottommost variable.

(c) In the previous steps the BDD size resulting from every variable swap has been recorded. Now the variable is moved back to the closest position among those positions which led to a minimal BDD size.

The algorithm is given in Figure 2.11. For $i < j$, a call sift_down$(i, j)$ moves a variable from level $i$ down to $j$ (the other procedure calls have similar semantics). We start with the largest level first, hence in lines (5), (7), and (9), $sl[i]$ denotes the number of the largest unprocessed level. In line (9) the tested condition is a check whether the way down is shorter than the way up. In that case, the algorithm goes down first. In line (16) the algorithm moves the variable back to a best position. This is usually done by following a sequence of recorded moves in reverse order.

Note that the run time of sifting increases, if the BDD becomes large in intermediate steps of the algorithm. To prevent high run times, moving a variable into a specific direction can be cancelled if the BDD size exceeds a certain limit, e.g. a limit of twice the size of the initial BDD.

### 2.4.8 Evaluation with BDDs

Besides the ability of BDDs to represent a Boolean function, BDDs can be also used to actually *implement an evaluation* of the function.

While evaluating functions with BDDs, one has to consider *paths* in BDDs, starting at one of the output nodes and ending at a terminal node. Sometimes also paths to an inner node are considered. Since BDDs essentially are DAGs, they inherit the usual standard notations considering paths in graphs. Paths in BDDs, as is common for graphs, are denoted as alternating sequence of nodes $v_i$ and edges $e_i$, i.e. $(v_1, e_1, \ldots, e_{k-1}, v_k)$. The length of a path $p$ is the number of non-terminal nodes occurring on $p$, denoted $\lambda(p)$. Next, an evaluation of a BDD with respect to an assignment is defined in operational terms.

```
sifting(BDD F, int n)
  proc
    sort level numbers by descending level sizes and store
    them in array sl;
    for i := 1 to n do
      if sl[i] = 1 then
        sift_down(i, n);
      else if sl[i] = n then
        sift_up(i, 1);
      else if (sl[i] − 1) > (n − sl[i]) then
        sift_down(i, n);
        sift_up(n, 1);
      else
        sift_up(i, 1);
        sift_down(1, n);
      end–if
      sift_back();
    end–for
  end–proc
```

*Figure 2.11.* Original sifting-algorithm.

An assignment $b = (b_1, \ldots, b_n) \in \mathbf{B}^n$ denotes the function

$$b\colon\ 2^{X_n} \rightarrow \mathbf{B}^n; x_i \mapsto b_i \quad (1 \leq i \leq n).$$

DEFINITION 2.26 *An* evaluation *of a BDD* $(\ldots, (V, E), O)$ *with respect to an assignment* $b = (b_1, \ldots, b_n) \in \mathbf{B}^n$ *starts at one of the output nodes in* $O$ *and traverses the path along the edges in* $E$ *which are chosen according to the values assigned to the variables by* $b$. *Thereby all variables which are not tested along the traversed path are ignored.*

*The evaluation is said to* reach *a node* $v \in V$ *if* $v$ *occurs on the traversed path. The evaluation is said to* stop *at node* $v$ *if* $v$ *is the last node of the traversed path.*

Due to the BDD semantics as a graph where a Shannon decomposition is carried out at each node, we have $f(b_1, b_2, \ldots, b_n) = 1$ iff the evaluation stops at **1** and $f(b_1, b_2, \ldots, b_n) = 0$ iff the evaluation stops at **0**.

EXAMPLE 2.27 *Consider the left BDD given in Figure 2.12. The evaluation for* $b = (0, 0, 1)$ *starts at the output node for* $f$ *which is the root node of the BDD. Assignment* $b$ *assigns* $x_1$ *to* 0, $x_2$ *to* 0, *and* $x_3$ *to* 1. *According to these values, the path along the corresponding edges labeled*

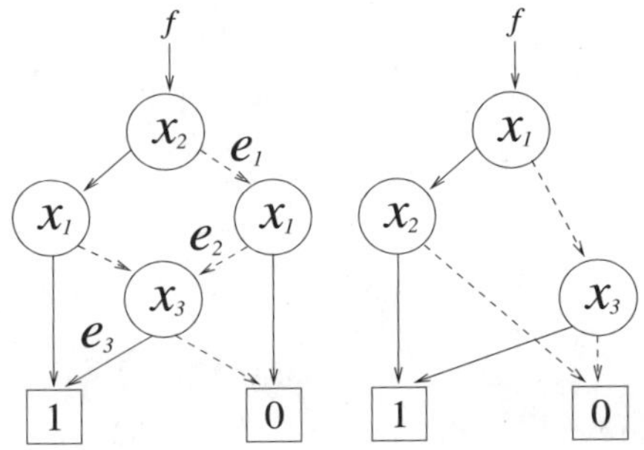


*Figure 2.12.* Two BDDs for $f\colon (x_1, x_2, x_3) \mapsto x_1 \cdot x_2 + \overline{x}_1 \cdot x_3$.

*$e_1$, $e_2$, and $e_3$ is chosen (1-edges are depicted with solid lines, 0-edges with dashed lines). This path finally reaches the terminal node labeled* 1*, and indeed the function value $f(0,0,1)$ equals* 1.

Note that the path along $e_1$, $e_2$, and $e_3$ is of maximal length three in the left BDD whereas the right BDD has a maximal path length of only two.

Let us consider a BDD respecting a variable ordering $\pi$. Sometimes we may choose to assign values to only the first few variables in the ordering $\pi$, thus considering a possibly shorter prefix $a = (b_1, \ldots, b_k)$ $(k \leq n)$ of a (full) assignment $b$. The operational semantics of an evaluation directly transfers to this situation in complete analogy. Evaluation of $a$ stops at a (possibly non-terminal) node $v$, representing the cofactor $f_{x_1=b_1,\ldots,x_k=b_k}$ for which we also write $f_a$.

## 2.4.9 Paths in BDDs

The optimization of BDDs with respect to different aspects of their paths will be subject to further discussion in Chapter 5. Thereby CEs (see Sections 2.4.2 and 2.4.3) will be explicitly considered to distinguish paths to zero from paths to one. Formally BDDs with CEs have to be represented by a edge-labeled graph in order to explicitly represent edges from a node $v$ to then($v$) and else($v$), respectively. This is necessary as the following example illustrates.

EXAMPLE 2.28 *Consider the projection function for variable $x_1$ shown in Figure 2.13. Both outgoing edges of node $v$ lead to the terminal* **1**. *Therefore they would correspond to a single edge in the graph structure without labels and the complement could not be properly associated to the edge leading to* else($v$)*. This ambiguity is removed using the edge-labeled graph that contains two edges.*

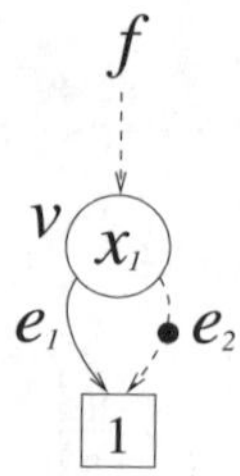


*Figure 2.13.* BDD with CEs for $f(x_1) = x_1$.

The following definitions are given for the more difficult case of BDDs with CEs only. For an edge $e \in E$ the attribute CE($e$) is true, iff $e$ is a CE. The edges $e_i$ occurring on a path in a BDD may be complemented or non-complemented. An (implicit) edge $e$ pointing to the root node has to be considered to represent a function. Such an implicit (possibly complemented) edge into the BDD is used to represent $f$ or $\overline{f}$, respectively. This edge is not explicitly denoted.

The predecessors of a node $w$ are split into those having a CE to $w$ and those having a regular edge to $w$, given by two sets (the sets are not always disjoint):

$$\begin{aligned} \mathcal{M}_1(w) &:= \{v : w \text{ can be reached from } v \text{ via a regular edge}\} \\ \mathcal{M}_0(w) &:= \{v : w \text{ can be reached from } v \text{ via a CE}\} \end{aligned}$$

Regarding the following terminology for paths in BDDs, it should be noted that for a BDD with CEs the implicit edge representing a function has to be considered.

DEFINITION 2.29

*Two paths*

$$\begin{aligned} p_1 &= (v_0, d_0, v_1, d_1, \ldots, d_{l-1}, v_l), \\ p_2 &= (w_0, e_0, w_1, e_1, \ldots, e_{l-1}, w_l) \end{aligned}$$

*with* $v_i, w_i \in V$ *and* $e_i, f_i \in E$ *are* identical, *iff*

$$\begin{aligned} \forall i \in \{0, \ldots, l\} \quad & v_i && = w_i, \\ \forall i \in \{0, \ldots, l-1\} \quad & d_i && = e_i, \\ \forall i \in \{0, \ldots, l-1\} \quad & \text{CE}(d_i) && = \text{CE}(e_i). \end{aligned}$$

*Otherwise the two paths are called* different.

DEFINITION 2.30 *A path* $p = (v_0, e_0, v_1, e_1, \ldots, e_{l-1}, v_l)$ *is called* complemented *iff it leads from* $v_0$ *to* $v_l$ *via an odd number of CEs, i.e.*

$$|\{e : (e \text{ is an edge in } p) \land \mathrm{CE}(e)\}| = 2i + 1$$

*for some* $i \in \mathbb{N}$. *Otherwise the path is called* regular.

DEFINITION 2.31 *A path* $p = (v_0, e_0, v_1, e_1, \ldots, e_{l-1}, v_l)$ *is called a* **1**-path *("one-path")* from $v_0$ *iff the path is regular and* $v_l = \mathbf{1}$. *A path* $p = (v_0, e_0, v_1, e_1, \ldots, e_{l-1}, v_l)$ *is called a* **0**-path *("zero-path")* from $v_0$ *iff the path is complemented and* $v_l = \mathbf{1}$.

#### 2.4.9.1 Number of Paths

NOTATION 1 *Let* $f$ *be a Boolean function.* $P_1(\mathrm{BDD}(f, \pi))$ *denotes the number of all different* **1**-*paths from any of the outputs to the terminal* **1** *with respect to the variable ordering* $\pi$. $P_0(\mathrm{BDD}(f, \pi))$ *denotes the number of all different* **0**-*paths from any of the outputs with respect to the variable ordering* $\pi$. *To denote the number of all paths in* $\mathrm{BDD}(f, \pi)$, *we use the short notation*

$$\alpha(\mathrm{BDD}(f, \pi)) = P_1(\mathrm{BDD}(f, \pi)) + P_0(\mathrm{BDD}(f, \pi)).$$

EXAMPLE 2.32 *The BDD for the well-known odd-parity or EXOR-function* $f\colon \mathbf{B}^n \to \mathbf{B}$; $(x_1, \ldots x_n) \mapsto x_1 \oplus \ldots \oplus x_n$ *is of linear size* $2n + 1$. *The corresponding BDD is shown in Figure 2.14. Nonetheless the number of paths is exponential in* $n$ *and, even worse, all BDDs representing the EXOR-function have a number of paths exponential in* $n$. *Due to the symmetry of the EXOR-function with respect to all variables, the structure of the BDD is independent of the variable ordering.*

Example 2.32 shows that the number of paths can grow exponentially in $n$, the number of input variables, even if the size of the BDD grows only linear in $n$.

When computing the number of paths in a BDD (see Section 5.1 and in particular Section 5.3) an exponential run time should be avoided. An appropriate formula for the number of paths in a BDD $F$ can be derived from the Shannon decomposition (Theorem 2.9). The following recurrent equation expresses the number of paths starting at a node $v$ and ending at a terminal node (this quantity denoted $\alpha(v)$):

$$\alpha(v) = \begin{cases} 1, & v \in \{\mathbf{1}, \mathbf{0}\} \\ \alpha(\mathrm{then}(v)) + \alpha(\mathrm{else}(v)), & \mathrm{else} \end{cases} \tag{2.2}$$

The formula simply states that $\alpha(v)$ is one if $v$ is a terminal node (there is one path of length zero from $v$ to $v$). Otherwise, the paths starting at

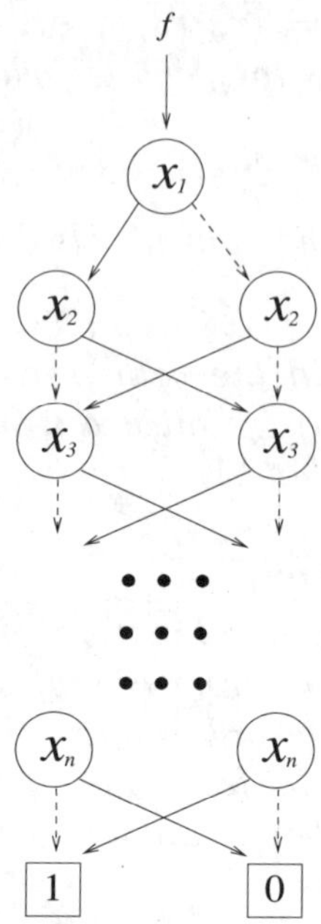


*Figure 2.14.* BDD for Example 2.32.

$v$ and ending at a terminal node can be partitioned into the paths via the 1-child and those via the 0-child, due to the Shannon decomposition carried out at each node. The sum of the number of paths in these two partitions yields the total number of paths.

It is straightforward to give an algorithm computing the number of paths during a graph traversal (an example of the recursive computation of the $\alpha$-values is illustrated later in Section 5.3). The number of paths in a shared BDD $F$ representing a Boolean multi-output function $f = (f_i)_{1 \leq i \leq m}$ can be computed as the sum of the $\alpha$-values of the output nodes representing the $m$ single-output functions:

$$\alpha(F) = \sum_{i=1}^{m} \alpha(o_i) \tag{2.3}$$

where $o_i$ is the output node representing $f_i$. Note that an output node $o_i$ might be used by multiple, functionally equivalent single-output functions, as circuits sometimes repeat an output signal several times. In other words, paths emerging from output nodes used for more than one function are counted several times accordingly.

#### 2.4.9.2 Expected Path Length

An important problem of VLSI CAD is *simulation-based verification*: here, a circuit is simulated to check whether the design fullfills its specification. The main difference of cycle-based *functional simulation* to classical simulation is that only values of output and latch ports are computed. Applications in this field repeatedly evaluate logic functions with different input vectors. This can be done using BDDs as described in Section 2.4.8.

During evaluation, a path is traversed. Thereby the algorithm moves from the start node of the path to the end of the path, following the respectiv edges. Pointers to the respective next node on the path must be dereferenced repeatedly. Hence, the evaluation time is linear in the length of the path. In BDD-based functional simulation, the average evaluation time (and hence the total simulation time) is proportional to the *Expected Path Length* (EPL) in the BDD. Next, a formal definition of the expected path length follows. EPL expresses the expected number of variable tests needed to evaluate a BDD with respect to an input assignment along a path from an output node to one of the terminal nodes, as explained and defined above in Section 2.4.8.

Let $F$ be a BDD and let $p_i$ be the $i$-th path in an enumeration of all paths from output nodes to one of the terminal nodes in $F$. Let $\mathrm{pr}(p_i)$ be the probability of an evaluation traversing path $p_i$. Then for the EPL of $F$, denoted $\epsilon(F)$, we have

$$\epsilon(F) = \sum_{i=1}^{\alpha(F)} \lambda(p_i) \cdot \mathrm{pr}(p_i). \tag{2.4}$$

In this formula, a path $p$ is weighted with the probability of being chosen during evaluation. Minimizing $\epsilon(F)$ means shortening the path lengths with a high probability, thus minimizing the expected path length or, in other words, the average evaluation time.

Equation (2.4) is not suitable for use by an efficient algorithm to compute EPL, as $\alpha(F)$ can grow exponentially in $n$, even if the size of the BDD only grows linear in $n$: a function and the respective BDD with this property has already been given in Example 2.32.

Next, a formula is given which is useful for computing EPL in a time proportional to the BDD size: the following equation expresses the expected number of variable tests for an evaluation starting from a node $v$ and ending at one of the terminal nodes (this quantity denoted $\epsilon(v)$).

Thereby the probability that a variable $x$ is assigned to a value $b \in \mathbf{B}$ is denoted by $\mathrm{pr}(x = b)$.

$$\epsilon(v) = \begin{cases} 0, & v \in \{\mathbf{1}, \mathbf{0}\} \\ 1 \quad + \quad \mathrm{pr}(\mathrm{var}(v) = 1) \cdot \epsilon(\mathrm{then}(v)), & \text{else} \\ \quad\;\; + \quad \mathrm{pr}(\mathrm{var}(v) = 0) \cdot \epsilon(\mathrm{else}(v)) & \end{cases}$$

This formula simply states that $\epsilon(v)$ is zero if $v$ is already a terminal node. Otherwise, evaluations starting from $v$ are either via the 1-child or the 0-child of $v$. Hence, $\epsilon(v)$ is built by

1) summing up the respective $\epsilon$-values of the child nodes of $v$ weighted with the probability of the respective child node being chosen, and

2) adding one since the expected length of all paths starting at $v$ must be one larger than that of the child nodes of $v$: this is due to the additional variable test at $v$.

Again it is straightforward to compute $\epsilon(F)$ during a graph traversal. An example of the recursive computation of the $\epsilon$-values is illustrated later in Section 5.2 by the left BDD in Figure 5.8. In analogy to Equation (2.3), the EPL for a shared BDD $F$ representing a Boolean multi-output function $f = (f_i)_{1 \leq i \leq m}$ can be computed by use of the $\epsilon$-values of the output nodes representing the $m$ single-output functions:

$$\epsilon(F) = \frac{1}{m} \sum_{i=1}^{m} \epsilon(o_i)$$

where $o_i$ is the output node representing $f_i$. Again, note that an output node $o_i$ might be used by multiple, functionally equivalent single-output functions.

#### 2.4.9.3 Average Path Length

In Section 5.3, optimization of BDDs with respect to the following aspect of their paths will be discussed. The motivation for this is the mapping of optimized BDDs into fast multiplexor-based circuits.

Let $F$ be a BDD and let $\lambda(F)$ denote the sum of the lengths of the paths in $F$. Again, let $p_i$ be the $i$-th path in an enumeration of all paths from output nodes to the terminal node in $F$. Then $\lambda(F)$ can be expressed as

$$\lambda(F) = \sum_{i=1}^{\alpha(F)} \lambda(p_i). \tag{2.5}$$

Equation (2.5) can be seen as a special instance of Equation (2.4) where all path probabilities $\text{pr}(p_i) = 1$.

The *Average Path Length* (APL) of $F$ is denoted $\overline{\lambda}(F)$ and is defined as

$$\overline{\lambda}(F) = \frac{\lambda(F)}{\alpha(F)}. \tag{2.6}$$

Equation (2.6) can be seen as a special case of Equation (2.4), defining the expected path length: the two equations coincide, if in Equation (2.4) the term $\text{pr}(p_i)$ is fixed to $\frac{1}{\alpha(F)}$ for each $p_i$.

A recurrent equation that is more useful for the efficient computation of the sum of the lengths of the paths is derived from the Shannon decomposition. This equation is given in the next result.

LEMMA 2.33 *For a node $v$ in a BDD, let $\lambda(v)$ denote the sum of the lengths of all paths starting at $v$ and ending at a terminal node. Then we have*

$$\lambda(v) = \begin{cases} 0, & v \in \{\mathbf{1}, \mathbf{0}\} \\ \lambda(\text{then}(v)) + \alpha(\text{then}(v)) & \\ \quad + \lambda(\text{else}(v)) + \alpha(\text{else}(v)), & \textit{else} \end{cases} \tag{2.7}$$

**Proof.** In the case of $v \in \{\mathbf{1}, \mathbf{0}\}$ there is nothing to show. Now let $v$ be an inner node. Let $p = (v, e_1, \text{then}(v), \ldots, t)$ be a path from $v$ via $\text{then}(v)$ onto a terminal node $t$. Further, let $p' = (\text{then}(v), \ldots, t)$ be a path coinciding with $p$ except that we start at $\text{then}(v)$ instead. It is $\lambda(p) = \lambda(p') + 1$, i.e. the length of every path via $\text{then}(v)$ is increased by one if started at $v$ instead. There are $\alpha(\text{then}(v))$ such paths. Consequently, the sum of the lengths of all paths from $v$ via $\text{then}(v)$ onto a terminal node must be $\lambda(\text{then}(v)) + \alpha(\text{then}(v))$. An analogous argument holds for paths via $\text{else}(v)$. The set of all paths starting from $v$ and ending at a terminal node can be partitioned into those via $\text{then}(v)$ and those via $\text{else}(v)$. But then the required result already follows. □

Similar to Equations (5.8) and (2.3), the sum of path lengths for a shared BDD $F$ representing a Boolean multi-output function $f = (f_i)_{1 \leq i \leq m}$ can be computed by use of the $\lambda$-values of the output nodes representing the $m$ single-output functions:

$$\lambda(F) = \sum_{i=1}^{m} \lambda(o_i)$$

# Chapter 3

# EXACT NODE MINIMIZATION

In this chapter, classical and recently published algorithms for the exact minimization of BDD size are given. The presentation starts with the method from [FS90]. Here, an important invariant for BDD level sizes has been observed. The level size is preserved under all variable movements that follow a certain restriction. The method then exploits this property and describes a significant reduction of the search space.

The review of classical approaches then continues with the method proposed in [ISY91]. This is the first method that is completely based on BDDs. Moreover, the pruning of the search space with the use of lower bounds has been introduced here. The lower bounds state restrictions on the BDD sizes that are achievable from a certain state of the computation. The bounds are tightened every time respective new information becomes available. Later works followed this *Branch and Bound* (B&B) paradigm for exact BDD minimization.

In [DDG00], a tighter lower bound has been suggested which drastically reduces the overall run time. By this larger functions can be handled, e.g. exact solutions for 32-bit adders have been computed successfully. A recent work [EGD03b] introduces an effective extension of the B&B technique that uses more than one lower bound in parallel. The additional lower bounds are obtained by a generalization of a lower bound known from VLSI design. The generalized bound can be also used for bottom-up construction of a minimized BDD. So the new approach is not restricted to a top-down construction like the approach of [DDG00]. Moreover, *combining* the two lower bounds yields a new lower bound that is used to exclude states earlier than in previous approaches, resulting in a further speed-up.

Recently, the change to another programming paradigm, ordered best-first search, i.e. the so-called $A^*$-algorithm as known from AI, has been suggested [EGD04a, EGD05]. This step has been prepared by [Ebe03] which suggests an efficient state expansion technique. Theory and implementation of the method are presented together with experimental results demonstrating the efficiency of the approach.

This chapter is structured as follows: Section 3.1 gives the B&B-based methods for exact BDD minimzation. In Section 3.1.1 classical B&B approaches to exact minimization of BDDs are reviewed. Next, a recent approach to exact BDD minimization is presented. It makes use of a new lower bound technique which is described in detail in Section 3.1.2: first, a generalization of a bound known from VLSI design is given which allows to construct the minimized BDD both bottom-up and top-down. Second, techniques to prune the state space at an early stage are described. In Section 3.1.3, a sketch of the code of this most recent B&B-based algorithm is given and new implementation techniques unique to the approach are explained. In Section 3.1.4, experimental results are presented.

Section 3.2 presents the $A^*$-based approaches to the optimization of BDD size. Some general background of state space search by $A^*$ is given in Section 3.2.1. This section also introduces some notations needed in later sections. The basics of state space search, heuristic search and the $A^*$-algorithm are outlined. In Section 3.2.2 it is explained, how the problem of finding an optimal variable ordering can be expressed in terms of minimum cost path search in a state space. It is outlined, how the generic $A^*$-algorithm can be used for this task. To ensure efficiency of $A^*$-based approaches, the heuristic function used must have an important property: this is the property of monotonicity. In Section 3.2.3, motivation and a formal proof of this property are given for the heuristic function chosen in the new approach. The recent algorithm presented here makes use of several techniques to improve the basic $A^*$-paradigm which are described in detail in Section 3.2.4. First, two techniques to combine $A^*$ and B&B are presented. Next, a technique to efficiently implement and maintain a basic data structure for $A^*$, a priority queue, is suggested. Then the technique to reduce the memory requirement of $A^*$ is presented which is based upon the reconstruction of paths. This section is finished with a new state expansion technique without expensive variable shifts. In Section 3.2.5 the approach is generalized from a pure top-down method to one being able to perform search both top-down and bottom-up, completing the presentation of the new algorithm. The last Section 3.2.6 presents experimental results from three test-suites.

## 3.1 Branch and Bound Algorithm

Typically, algorithms for exact BDD minimization have to accomplish search spaces of large, i.e. exponential size. This clearly holds for a naive brute-force approach which examines all of the $n!$ orderings. An observation in [FS90] can be used to reduce this size significantly. However, since it is NP-complete to decide whether the number of nodes of a given BDD can be improved by variable reordering [BW96], the run time complexity of all exact minimization algorithms presented so far has been significantly higher (i.e. exponential) than that of mere "rules of thumb" to find a "good" variable ordering, i.e. heuristics.

Traditionally, this problem has been tackled by B&B methods which prune the search space with lower and upper bounds to reduce run time.

This section first gives a description of the classical B&B methods. The presentation of the basic ideas behind them is supported by giving pseudo code for the most important methods. The stages of development and the differences between the respective algorithms are clarified.

Next, special emphasis is put on the latest developments: the most recently published exact B&B technique for determining an optimal variable ordering is presented. In contrast to all previous approaches that only considered one lower bound, this method makes use of a combination of three bounds and by this avoids unnecessary computations. The lower bounds are derived by generalization of a lower bound known from VLSI design. They allow to build the BDD either top-down or bottom-up.

### 3.1.1 Classical Approach

In 1987 and 1990, respectively, an approach was presented working on truth tables [FS90] where the number of considered variable orderings has been reduced to $2^n$, a significant improvement over the "naive" approach considering all $n!$ variable orderings. The reduction of the search space was based on the following observation:

LEMMA 3.1 *Let* $f\colon \mathbf{B}^n \to \mathbf{B}^m$, $I \subseteq X_n$, $k = |I|$, *and* $x_i \in I$. *Then there exists a constant* $c$ *such that* $\text{label}(\text{BDD}(f, \pi), x_i) = c$ *for each* $\pi \in \Pi(I)$ *with* $\pi(k) = x_i$.

More informally, this lemma can be rephrased as follows: the number of nodes in a level is constant if the corresponding variable is fixed in the variable ordering and no variables from the lower and upper part are exchanged. Note that this holds independently of the ordering of the variables in the upper and lower part of the BDD. Using this invariant, the approach of [FS90] is capable of excluding many variable orderings from consideration as they lead to equivalent level sizes.

(1) **compute_optimal_ordering(BDD** $F$**, int** $n$**)**
(2)     **proc**
(3)         min_cost$(\emptyset, \emptyset) := 0$;
(4)         $\pi_\emptyset :=$ an arbitrary initial order;
(5)         **for** $k := 1$ **to** $n$ **do**
(6)             **for each** $k$-element subset $I \subseteq \{1, 2, \ldots, n\}$ **do**
(7)                 Compute min_cost$(I, I)$ and $\pi_I$ using the terms
                    min_cost$(I \setminus \{x_i\}, I \setminus \{x_i\})$ and
                    $\pi_{I\setminus\{x_i\}} \quad (x_i \in X_n \setminus I)$;
(8)             **end–for**
(9)         **end–for**
(10)    **end–proc**

*Figure 3.1.* The algorithm of Friedman and Supowit.

Next, two new ideas have been introduced in [ISY91]: the use of BDDs instead of truth tables and the use of *upper and lower bounds* on BDD sizes. The idea is to skip examination of BDDs as soon as a lower bound on the sizes of BDDs achievable from that BDD already exceeds an upper bound for the minimal size. The problem now is seen as a *search problem* in a *state space* where states are subsets of $X_n$. A B&B technique is used to skip large parts of the state space.

This approach has been further enhanced in [JKS93]: the search space is reduced if the function has symmetric variables (see Definition 2.3) and a better lower bound is used. The algorithm in [JKS93] also is the first that is implemented using hash tables.

In [DDG00], a lower bound known from VLSI design has been adapted for exact minimization. This method also is the first one which applies a top-down approach (all previous methods were applied bottom-up). The approach showed a speed-up factor of up to 400 when compared to [JKS93].

Next the basic minimization algorithm following the framework from [FS90] is described. Suppose the BDD for a multi-output function $f\colon \mathbf{B}^n \to \mathbf{B}^m$ is given.

In brief, the optimal variable ordering is computed iteratively by computing for increasing $k$'s min_cost$(I, I)$ for each $k$-element subset $I$ of $X_n$, until $k = n$: then, the BDD has a variable ordering yielding a BDD size of min_cost$(X_n, X_n)$. This is an optimal variable ordering.

A sketch of the basic algorithm is shown in Figure 3.1. The computation in line (7) is based on a recurrent interrelation that relates min_cost$(I, I)$ and $\pi_I$ to min_cost$(I \setminus \{x_i\}, I \setminus \{x_i\})$ and $\pi_{I\setminus\{x_i\}}$ where $\pi_I$ is a variable ordering such that

$$\text{label}(\text{BDD}(f, \pi_I), I) = \text{min\_cost}(I, I).$$

The same framework also has been used in [ISY91, JKS93, DDG00, EGD03b]. All recent approaches perform BDD operations and make use of hash tables as the underlying data structure. Next we follow the presentation of the exact BDD minimization algorithm in [DDG00].

As input, the algorithm receives a BDD $F$ representing a Boolean multi-output function $f$: $\mathbf{B}^n \to \mathbf{B}^m$ and $n$. At step $k$ of the algorithm given in Figure 3.2, a state $I$ with $|I| = k - 1$ is retrieved from a hash table (which holds all states of the previous step $k - 1$). The algorithm now generates transitions[1]

$$I \xrightarrow{x_i} I \cup \{x_i\} =: I' \quad (x_i \in X_n \setminus I),$$

see line (10) in Figure 3.2. Note that each symmetry set is only considered once (see the call of macro CHECK-SYMMETRY in line (9) in Figure 3.2. The macro is given in Figure 3.4). This restricts the search space if the function is partially symmetric, an optimization which has also been used in [JKS93]. The subject is to compute min_cost$(I', I')$ for each successor $I'$. This is done by a gradual schema of continuous minimum updates, using a reccurrent equation following [FS90]:

$$\begin{aligned}&\text{min\_cost}(I', I')\\ =\ &\min_{x_j \in I'} (\text{min\_cost}(I' \setminus \{x_j\}, I' \setminus \{x_j\}) + \text{label}(\text{BDD}(f, \pi_j), x_j))\end{aligned}$$

where $\pi_j$ is a variable ordering contained in $\Pi(I' \setminus \{x_j\})$ such that $\pi_j(|I'|) = x_j$.

This recurrence is a consequence of Lemma 3.1, which has been given at the beginning of this section.

The code implementing the update schema is embraced by the for-loop starting at line (6) in Figure 3.2: *minguess*[hash$(I)$] is updated until it reaches min_cost$(I, I)$. Note that within the loop we always have $k = |I| + 1$. The terms min_cost$(I' \setminus \{x_j\}, I' \setminus \{x_j\})$ have been saved in the hash table in the previous step $k - 1$ since $|I' \setminus \{x_j\}| = k - 1$ (see line (7) in Figure 3.5). Their values are simply retrieved from the hash table. The only terms still left to compute are label(BDD$(f, \pi_j), x_j)$ for each $x_j \in I'$ (see line (15) in Figure 3.2).

[1] In the approach of [EGD03a], the transition is generated only if $I'$ has not already been excluded (see Section 3.1.2.2).

```
(1)   compute_optimal_ordering(BDD F, int n)
(2)     proc
(3)       INIT-BB
(4)       for k := 1 to n do
(5)         next_states := ∅;
(6)         for each I ∈ states do
(7)           reconstruct(F, π[hash(I)]);
(8)           for each x_i ∈ X_n \ I do
(9)             CHECK-SYMMETRY
(10)            I' := I ∪ {x_i};
(11)            if I' ∉ next_states then
(12)              minguess[hash(I')] := ∞;
(13)            end–if
(14)            shift x_i to level |I| + 1;
(15)            newcost :=
                label(F, |I| + 1) + minguess[hash(I)];
(16)            UPDATE-STATE-DATA-FizZ
(17)            UNDO-SHIFT
(18)            end–if
(19)          end–for
(20)        end–for
(21)        exclude all states I' in next_states with
            lower_bound[hash(I')] ≥ upper_bound;
(22)        states := next_states;
(23)      end–for
(24)      reconstruct the ordering of upper_bound;
(25)    end–proc
```

*Figure 3.2.* The algorithm FizZ.

For $I := I' \setminus \{x_j\}$, let $\pi_I$ be a variable ordering such that $\pi_I \in \Pi(I)$. Then BDD$(f, \pi_j)$ can be constructed from BDD$(f, \pi_I)$ by shifting variable $x_j$ to the $(|I| + 1)$-th level (see line (14) in Figure 3.2). That way, the minimized BDD is built top-down, starting with the first level and, as $k$ increases, repeatedly adding another level below the current level. In [DDG00], the variable ordering $\pi_{I \cup \{x_j\}}$ resulting by this variable shift of $x_j$ is saved in a hash table for later steps. Since $\pi_{I \cup \{x_j\}}$ again has the desired property, i.e. it is contained in $\Pi(I \cup \{x_j\})$, this ordering can be used in the next step like $\pi_I$ was used for $I$.

To reduce run time and memory requirement, in Figure 3.2 the variable shift of line (14) is undone in line (17), if this shift caused an in-

**INIT-BB**<br>
  **macro**<br>
    $minguess[\text{hash}(\emptyset)] := 0$;<br>
    $\pi[\text{hash}(\emptyset)] :=$ an arbitrary initial order;<br>
    $states[\text{hash}(\emptyset)] := \emptyset$;<br>
  **end–macro**

*Figure 3.3.* Initialization for the B&B-based approach.

**CHECK-SYMMETRY**<br>
  **macro**<br>
    **if** $x_i$ is not top of ((symmetry group of $x_i$) $\cap (X_n \setminus I)$)<br>
    **then**<br>
      continue with for-loop;<br>
    **end–if**<br>
  **end–macro**

*Figure 3.4.* Checking for the symmetry of variables.

**UPDATE-STATE-DATA-FizZ**<br>
  **macro**<br>
    **if** $I' \notin next\_states$ **or** $newcost < minguess[\text{hash}(I')]$<br>
    **then**<br>
      $minguess[\text{hash}(I')] := newcost$;<br>
      $\pi[\text{hash}(I')] :=$ current ordering;<br>
      $upper\_bound :=$ update_upper_bound();<br>
      $next\_states[\text{hash}(I')] := I'$;<br>
      $lower\_bound[\text{hash}(I')] :=$ compute_lower_bound($I'$);<br>
    **end–if**<br>
  **end–macro**

*Figure 3.5.* Updating the state data in algorithm FizZ.

crease of the BDD size exceeding a threshold (see macro UNDO-SHIFT in Figure 3.6). This is another improvement compared to the previous approach from [JKS93].

At the end of step $k$, all states whose lower bound exceeds or equals the current upper bound, are excluded (see line (21) in Figure 3.2).

```
UNDO-SHIFT
    macro
        if (|BDD(f, current ordering)| >
        1.5 · (size before shifting x_i)) then
            undo shift of x_i;
        end–if
    end–macro
```

*Figure 3.6.* Undoing the variable shifts in algorithm FizZ.

In fact the BDD can also be built bottom-up following the same outline. Note that the approaches from [ISY91, JKS93] are bottom-up *only* whereas the approach from [DDG00] is top-down *only.*

### 3.1.2 Lower Bound Technique

In this section, the basic lower bound techniques of [DDG00] and the latest B&B-approach from [EGD03b] are given.

#### 3.1.2.1 A Generalized Lower Bound

Before the generalized lower bound is presented, a brief review of the lower bound used in [DDG00] is given which is an adaptation of a lower bound known from VLSI design, proposed by [Bry91].

DEFINITION 3.2 *Let $F$ be a BDD. For $k > 0$, let* $\mathrm{ref}(F,k)$ *denote the set of nodes in levels $k+1,\ldots,n$ of $F$ referenced directly from the nodes in levels $1,\ldots,k$ of $F$. If a node has no direct, i.e. only external references, it is* not *contained in* $\mathrm{ref}(F,k)$. *Let* $\mathrm{ref}(F,0)$ *denote the set of externally referenced nodes, i.e. the set of nodes which represent user functions. The set* $\mathrm{ref}(F,0)$ *is equal to the set of output nodes in $F$.*

*Formally, we have*

$$\mathrm{ref}_n\colon \{F \mid F \textit{ is a BDD over } X_n\} \times \mathbb{N} \to 2^{\{v\in V \mid (\ldots,(V,E),\ldots) \textit{ is a BDD}\}};$$

$$\mathrm{ref}_n(F,k) = \begin{cases} O, & k = 0 \\ \mathcal{R}_n(F, F_k^1) \cap F_n^{k+1}, & k > 0 \end{cases}$$

*where $F = (\ldots,\ldots,O)$ and*

$$\mathcal{R}_n\colon \{F \mid F \textit{ is a BDD over } X_n\} \times 2^{\{v\in V \mid (\ldots,(V,E),\ldots) \textit{ is a BDD}\}} \to 2^{\{v\in V \mid (\ldots,(V,E),\ldots) \textit{ is a BDD}\}};$$

$$\mathcal{R}_n(F,N) = \{v \mid (u,v) \in E \textit{ where } F = (\ldots,(V,E),\ldots) \textit{ and } u \in N\}.$$

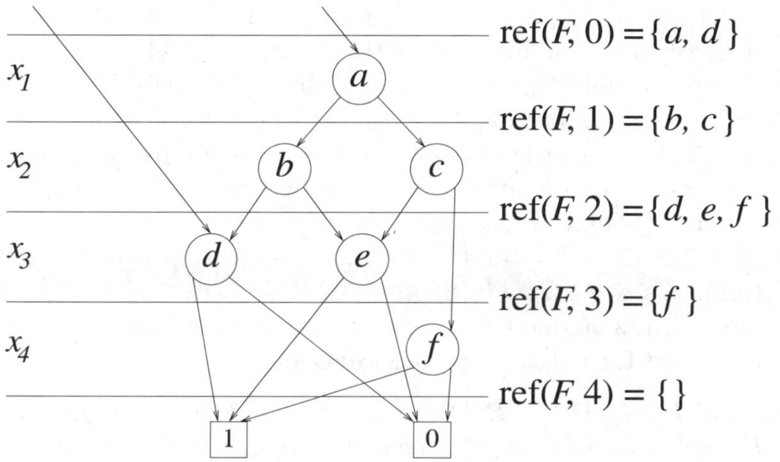


*Figure 3.7.* BDD for Example 3.3.

As before, the function ref usually is denoted omitting the subscript $n$ since $n$ normally is given by the context.

EXAMPLE 3.3 *Consider the BDD F given in Figure 3.7. Note that the node labels are node identifiers and not variables (which are annotated at the levels instead). The two outputs are represented by nodes a and d, thus* $\text{ref}(F, 0) = \{a, d\}$. *This set as well as the other sets are given in the figure.*

LEMMA 3.4 *Let* $f\colon \mathbf{B}^n \to \mathbf{B}^m$ *be a multi-output function. Let* $(L, R)$ *be a partition of* $X_n$, $\pi \in \Pi(L)$ *and* $F := \text{BDD}(f, \pi)$. *If* $|\text{ref}(F, |L|)| = c$, *then each BDD with a variable ordering in* $\Pi(L)$ *representing* $f$ *has at least* $c$ *nodes in levels* $|L| + 1, \ldots, n$.

A proof can be found in [Bry91, DDG00]. In exact minimization, this argument is used to obtain a lower bound based on $\text{ref}(F, |I|)$ at each step, a state $I \subseteq X_n$ is considered. Suppose that the minimized BDD $F$ is constructed top-down. Further assume that, when computing the lower bound for a multi-output function $f\colon \mathbf{B}^n \to \mathbf{B}^m$, the minimal number $\text{min\_cost}(I, I)$ of nodes in levels $1, \ldots, |I|$ is already known. Let $c_\alpha = |\text{ref}(F, |I|)|$. Then by Lemma 3.4, the lower bound can be computed as

$$l\_b_\alpha = \text{min\_cost}(I, I) + \max\{c_\alpha + r\_lower, n - |I|\} + 1. \qquad (3.1)$$

In order not to count some output nodes twice, *r_lower* is the number of output nodes in levels $|I| + 1, \ldots, n$ not already representing a node in $\text{ref}(F, |I|)$ and $n - |I|$ is the number of variables in $X_n \setminus I$ since there will be at least one node for each of these variables. The constant node is always needed (note that we here consider BDDs *with* CEs).

A detailed analysis reveals that $l\_b_\alpha$ can be defined as a *function* in $f^{(n)}$ and $I \subseteq X_n$. In order to present the most important issues first, not to get lost in details at this stage of presentation and to obtain a shorter, more readable definition, this lower bound is not yet defined as a formal function of $I$ (and $F$ or $f$, respectively). However, a more detailed analysis will follow in Section 3.2.2.

The next results enable us to generalize this lower bound such that it can be also used for bottom-up construction.

LEMMA 3.5 *Let $f$: $\mathbf{B}^n \to \mathbf{B}^m$ be a multi-output function. Let $I \subseteq X_n$, $\pi \in \Pi(I)$ and $F := \text{BDD}(f, \pi)$. Then we have $|\text{ref}(F, |I|)| - r\_upper \leq |F^1_{|I|}|$ where $r\_upper$ is the number of output nodes in levels $1, \ldots, |I|$ of $F$.*

**Proof.** The idea is to calculate the minimal number of nodes needed to connect the output nodes with the nodes in $\text{ref}(F, |I|)$.

For every node $v \in |F^1_{|I|}|$ different from the output nodes[2], there must be at least one outgoing edge of a node in $|F^1_{|I|}|$ leading to $v$. On the other hand, there are $2 \cdot |F^1_{|I|}|$ outgoing edges of nodes in $|F^1_{|I|}|$ in total. Hence, the number of edges *not* leading to a node in $|F^1_{|I|}|$ must be less or equal to

$$2 \cdot \left|F^1_{|I|}\right| - \left(\left|F^1_{|I|}\right| - r\_upper\right) = \left|F^1_{|I|}\right| + r\_upper.$$

This number equals the number of references to nodes in $\text{ref}(F, |I|)$, thus we have $|\text{ref}(F, |I|)| \leq |F^1_{|I|}| + r\_upper$ or

$$|\text{ref}(F, |I|)| - r\_upper \leq \left|F^1_{|I|}\right|, \tag{3.2}$$

completing the proof. □

The main result yielding the new lower bound is a corollary of Lemma 3.5:

COROLLARY 3.6 *Let $f$: $\mathbf{B}^n \to \mathbf{B}^m$ be a multi-output function. Let $I \subseteq X_n$, $\pi \in \Pi(I)$ and $F := \text{BDD}(f, \pi)$. Then we have $|\text{ref}(F, |I|)| - r\_upper \leq \text{min\_cost}(I, I)$ where $r\_upper$ is the number of output nodes in levels $1, \ldots, |I|$ of $F$.*

---

[2]Note that these nodes cannot be root nodes since any root in a BDD is an output node.

**Proof.** Inequality (3.2) stated in Lemma 3.5 holds for every BDD $F$ with a variable ordering in $\Pi(I)$. In particular, this also holds for a BDD $\mathcal{F}$ which is optimal for $I$, i.e. with a variable ordering such that $|\mathcal{F}^1_{|I|}| = \text{min\_cost}(I, I)$. It is $|\text{ref}(\mathcal{F}, |I|)| - r\_upper \leq \text{min\_cost}(I, I)$. Now we claim that the value for the left side of Inequality (3.2) remains the same for all BDDs $F$ with a variable ordering in $\Pi(I)$:

By the BDD semantics in Definition 2.14, the number of output nodes $r\_upper$ must be the same for every such ordering since otherwise $F$ would not represent $f$. The nodes in $\text{ref}(F, |I|)$ represent true cofactors of $f$ in all variables in $I$[3].

Their number remains the same for every variable ordering in $\Pi(I)$ since, again, otherwise $F$ would not represent $f$.

Thus indeed the left side does not change. Consequently we have

$$|\text{ref}(F, |I|)| - r\_upper = |\text{ref}(\mathcal{F}, |I|)| - r\_upper \leq \text{min\_cost}(I, I)$$

for all BDDs $F$ respecting a variable ordering $\pi \in \Pi(I)$. □

Suppose now the minimized BDD is constructed bottom-up. The BDD must respect a variable ordering in $\Pi(X_n \setminus I)$. Let $c_\omega = |\text{ref}(F, n - |I|)|$. Then a lower bound can be computed as

$$l\_b_\omega = \text{min\_cost}(I, X_n \setminus I) + \max\{c_\omega - r\_upper, n - |I|\} + 1$$

where $r\_upper$ is the number of output nodes in levels $1, \ldots, n-|I|$. This lower bound holds since, by Lemma 3.6, $c_\omega - r\_upper$ is a lower bound for $\text{min\_cost}(X_n \setminus I, X_n \setminus I)$ which is the minimal size of the upper part of the BDD.

Yet, the term $c_\omega$ corresponds to $c_\alpha$ in the lower bound $l\_b_\alpha$. But this time the number of direct references from the upper to the lower part is not used as a lower bound on the number of nodes in the *lower* part. Instead it serves as a lower bound on the number of nodes in the *upper part*. This is possible by Lemma 3.6 since a key result of this lemma is that the term $\text{ref}(F, k)$, besides expressing a lower bound on the number of nodes in the last $n - k$ levels, also can be used to express a lower bound on the number of nodes in the *first* $k$ levels as well. This appears counter-intuitive at first sight, and indeed this result is not trivial.

---

[3]This follows from the BDD semantics (see Definition 2.14 in Section 2.4), and from Theorem 2.17. Since this can be seen intuitively, we do not give a detailed proof here. However, a full and detailed proof can be found in Section 3.2.2, Lemma 3.10, Equation (3.5). This proof does not use previous results. Hence, this postponement of a more detailed proof is considered to support a better readability rather than being harmful in any way.

The lower bound $l\_b_\omega$, like before $l\_b_\alpha$, is a function in $f^{(n)}$ and $I \subseteq X_n$. This will be shown in detail in Section 3.2.5. Examples for the computation of $l\_b_\alpha$ and $l\_b_\omega$ are given in Example 3.7 and Example 3.8.

EXAMPLE 3.7 *In Figure 3.8, an example for the computation of the lower bound stated in Equation 3.4 is given. The (single-output) functions represented by the BDD are*

$$\begin{aligned} f_1 &= x_3, \\ f_2 &= x_1 \cdot x_2 \cdot x_3 + \overline{x}_1 \cdot x_3 + \overline{x}_1 \cdot \overline{x}_3 \cdot x_4, \\ f_3 &= x_4. \end{aligned}$$

*In the upper part of the BDD, consisting of the first two levels, a minimal size has already been achieved: we need at least one BDD node on every level since the function represented essentially depends both on $x_1$ and on $x_2$. The two nodes in the lower part with direct references from nodes of the upper part are lighter shaded. In the example, r_lower is (only) one: the output node pointed to by the edge labeled $f_1$ also has a direct reference from a node in the upper part. Hence it is not counted again. The (darker shaded) output node pointed to by $f_3$ however, has no direct reference from a node in the upper part and hence it is counted. Hence we have*

$$\begin{aligned} l\_b_\alpha &= \text{min\_cost}(\{x_1, x_2\}, \{x_1, x_2\}) + \max\{|\text{ref}(F, 2)| + 1, 4 - 2\} + 1 \\ &= 2 + \max\{3, 2\} + 1 \\ &= 2 + 3 + 1 \\ &= 6. \end{aligned}$$

*The lower bound yields six nodes. In the diagram, there is one (constant) node more, i.e. a total of seven nodes: this is due to the BDD being a BDD without CEs. Note that we easily can modify the lower bound accordingly, always adding 2 instead of adding 1 for the one constant node in BDDs with CEs.*

EXAMPLE 3.8 *In Figure 3.9, an example for the computation of the lower bound stated in Corollary 3.6 is given. The BDD represents the same functions as in Example 3.7. In the lower part of the BDD, consisting of the last two levels, a minimal size already has been achieved: to see this, consider the two BDDs in Figure 3.18 for the two possible orderings of the variables $x_3$ and $x_4$ in the lower part. Both BDDs show three nodes in their lower part, hence this size of three is minimal. In*

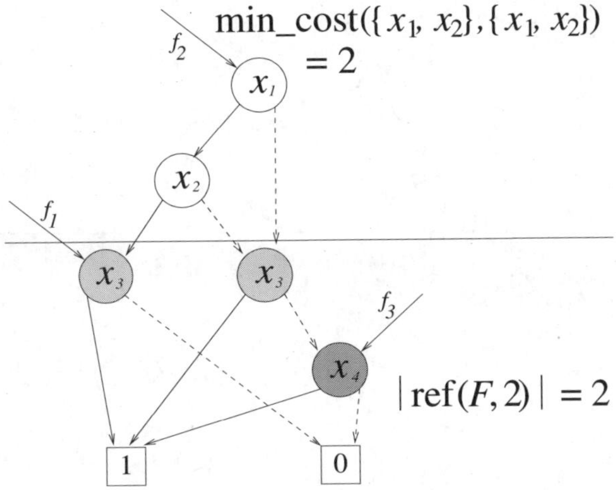


*Figure 3.8.* BDD for Example 3.7.

*the example, r_upper is one since we have one output node in the upper part (this is the node pointed to by the edge labeled $f_2$).*

*Again, the two nodes with direct references from nodes in the upper part are shaded. Now we have*

$$\begin{aligned} l\_b_\omega &= \text{min\_cost}(\{x_3, x_4\}, \{x_1, x_3\}) + \max\{|\text{ref}(F,2)| - 1, 2\} + 1 \\ &= 3 + \max\{1, 2\} + 1 \\ &= 3 + 2 + 1 \\ &= 6. \end{aligned}$$

*The lower bound again yields six nodes (the same remark for CEs holds as in Example 3.7). Note that only one of the two nodes in the upper part is "predicted" by the term* $\text{ref}(F,2) - 1 = 2 - 1 = 1$*). Computing the maximum with the term* $n - 2 = 4 - 2 = 2$ *however again yields the exact number of nodes.*

In the next section, the introduced lower bounds $l\_b_\alpha$ and $l\_b_\omega$ will be combined to a new lower bound which is used to exclude states at an early stage of the algorithm.

#### 3.1.2.2 Early Pruning

Two techniques to exclude states from further examination *as early as possible* are described. These techniques save a significant number of transitions from one state to another. As transitions involve vari-

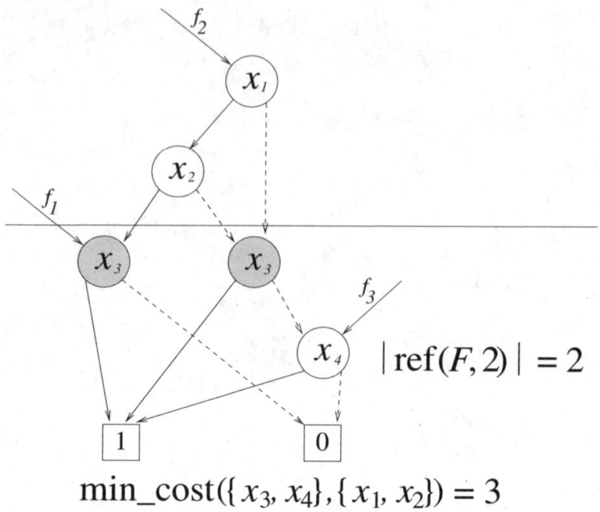


*Figure 3.9.* BDD for Example 3.8.

able shifts at high computational cost, this is a significant gain for the discussed algorithm.

The situation in which the techniques are applied is briefly described for the case of a top-down approach (all techniques directly transfer to the case of a bottom-up approach). In step $k$, the algorithm expands all states $I$ with $|I| = k - 1$ that have not been excluded in the last step, to all possible successors. Successor states are being revisited frequently in the progress of step $k$ since many distinct states $I$, $I'$ have successors in common. When repeatedly revisiting such a successor $J$, $\mathrm{min\_cost}(J, J)$ is gradually computed by continously updating the previous smallest number of nodes labeled with a variable in $J$. This number reaches $\mathrm{min\_cost}(J, J)$ at the end of step $k$.

It is desirable to

1) avoid transitions to successors which are already known *not* to contribute to the actualization of the smallest node number,

2) find a means which allows to test every successor for a possible exclusion right after it was generated: in this way, unnecessary repeated movements to successor states can be avoided.

Regarding the first point 1), consider a transition $I \xrightarrow{x_i} I \cup \{x_i\}$ where $I \cup \{x_i\}$ is a state that has already been visited before.

Let $F_I$ be the BDD for $I$. Since state $I$ was processed in the previous step, we already have label$(F_I, I)$ = min_cost$(I, I)$. The potential successor state would be built by e.g. shifting variable $x_i$ to level $|I| + 1$, resulting in a new BDD $F'$. A lower bound on label$(F', I \cup \{x_i\})$ is

$$l\_b_{cost} = \text{min\_cost}(I, I) + 1$$

since there will be at least one node labeled with the variable $x_i$. Note, that this lower bound[4] can be computed in constant time. If this lower bound is not smaller than the previous smallest number of nodes labeled with a variable in $I \cup \{x_i\}$, neither is the exact value label$(F', I \cup \{x_i\})$. In this case this transition is skipped since revisiting this state does not contribute to the actualization of its previous minimum.

Regarding the second point 2): at an early stage of step $k$ min_cost$(I, I)$ cannot be used to compute a lower bound since the exact value of the minimum is not known until step $k$ finishes. Therefore min_cost$(I, I)$ must be estimated.

This can be done both efficiently and effective by combining $l\_b_\alpha$ and $l\_b_\omega$. The idea is to estimate the (yet) unknown exact part of the lower bound, i.e. min_cost$(I, I)$, with the according estimated part of the opposite lower bound.

The next definition uses a partition $(L, R)$ of $X_n$ rather than states $I \subseteq X_n$ such that it applies for both the top-down and the bottom-up approach of exact minimization. In case of starting the minimization from above, we have $I = L$ at a step considering state $I$ (starting from below, we have $I = R$ and $L = X_n \setminus I$).

A lower bound for the minimal size of the BDD achievable from a partition $(L, R)$ can be computed as follows. Let $F$ be the considered BDD and $c_{\alpha\omega} = |\text{ref}(F, |L|)|$. Let

$$l\_b_{combined} = \max\{c_{\alpha\omega} + r\_lower, |R|\} + 1 + \max\{c_{\alpha\omega} - r\_upper, |L|\}$$

where $r\_lower$ is the number of output nodes in levels $|L| + 1, \ldots, n$ not already representing a node in ref$(F, |L|)$ and $r\_upper$ is the number of output nodes in levels $1, \ldots, |L|$.

All states already excluded "early" by this lower bound are marked by the method using this technique. Transitions leading to such a marked state are not followed by the algorithm, saving again the computational cost for a variable shift.

This combined lower bound is a function in $f^{(n)}$ and $L \subseteq X_n$. This property is essentially inherited from the lower bounds combined, $l\_b_\alpha$ and $l\_b_\omega$.

[4]More precisely, this lower bound can be expressed as a function in $f^{(n)}$ and $I \subseteq X_n$.

### 3.1.3 Algorithm

In this section the implementation techniques used in the latest published B&B-approach are described. A sketch of the algorithm called JANUS is given in Figure 3.10. The algorithm is based on the framework of FizZ which was given in Figure 3.2. The innovation of JANUS over FizZ is the use of more than one lower bound in parallel and several unique implementation techniques.

Differences between the two algorithms start at line (7) in Figure 3.10: the subroutine for BDD reconstruction of algorithm JANUS requires the additional parameter $|I|$. The improved approach to BDD reconstruction is described in Section 3.1.3.2. In line (14) of Figure 3.10 states can be tested for exclusion from the state space search already in an earlier stage of progress of the algorithm (see the code of macro CHECK-EXCLUDED-B&B in Figure 3.11).

The actual exclusion of a state following the "early pruning" techniques is done in line (10) in Figure 3.12. The techniques used here have been described in Section 3.1.2.2.

Another difference of macros UPDATE-STATE-JANUS and UPDATE-STATE-FizZ (the latter is given in Figure 3.5) is that the lower bound is only computed once when visiting a state the first time (see lines (7) and (9) in Figure 3.12). This improvement will be explained in Section 3.1.3.1.

Next, some new unique implementation techniques applied in the presented approach are described.

#### 3.1.3.1 Lower Bound Computation

To determine the lower bounds introduced in Section 3.1.2.1 for the current BDD $F$, $|\mathrm{ref}(F,k)|$ must be computed where $k = |I|$ for $l\_b_\alpha$ and $k = n - |I|$ for $l\_b_\omega$.

Computation of $|\mathrm{ref}(F,k)|$ is done with two different methods: one touches only the nodes in the upper part, i.e. in the first $k$ levels, the other touches only the nodes in the lower part, i.e. in the last $n-k$ levels. If the size of the upper part is smaller than that of the lower part, the first routine is called (since this is more promising in terms of expected run time) and vice versa. Since commonly used BDD packages keep and continously update the level sizes in dedicated variables, the time needed to determine the size of the upper and lower part is very small, i.e. it can be neglected.

It is sufficient to calculate the lower bounds only once the first time a state $I$ is encountered: assuming $|I| = k$, $\pi \in \Pi(I)$, $F := \mathrm{BDD}(f,\pi)$ the nodes in $\mathrm{ref}(F,k)$ represent the true cofactors of the single-output

```
(1)  compute_optimal_ordering(BDD F, int n)
(2)    proc
(3)      INIT-BB
(4)      for k := 1 to n do
(5)        next_states := ∅;
(6)        for each I ∈ states do
(7)          reconstruct(F, π[hash(I)], |I|);
(8)          for each x_i ∈ X_n \ I do
(9)            CHECK-SYMMETRY
(10)           I' := I ∪ {x_i};
(11)           if I' ∉ next_states then
(12)             minguess[hash(I')] := ∞;
(13)           end–if
(14)           CHECK-EXCLUDED-B&B
(15)           shift x_i to level |I| + 1;
(16)           newcost :=
               label(F, |I| + 1) + minguess[hash(I)];
(17)           UPDATE-STATE-DATA-JANUS
(18)           UNDO-SHIFT
(19)           end–if
(20)         end–for
(21)       end–for
(22)       exclude all states I' in next_states with
           lower_bound[hash(I')] ≥ upper_bound;
(23)       states := next_states;
(24)     end–for
(25)     reconstruct the ordering of upper_bound;
(26)   end–proc
```

*Figure 3.10.* The B&B algorithm JANUS.

functions $(f_i)_{1 \le i \le m}$ in all variables in $I$. A formal proof and further discussion is given later in Section 3.2.2 when discussing an exact BDD minimization algorithm based on the generic $A^*$-algorithm (see Lemma 3.10). The number of nodes representing such a cofactor is determined by $f$ and $I = \{\pi(1), \pi(2), \ldots, \pi(k)\}$ only and hence does not depend on which variable ordering in $\pi \in \Pi(I)$ is used for $F$. The algorithm gains from this as follows: the invariant terms of the lower bounds are computed separately from min_cost$(I, I)$ (and thus they are computed only once). This optimization is applied in the lines (7) and (9) of macro

```
CHECK-EXCLUDED-B&B
  macro
    if I' already excluded or lb_cost_I' ≥ minguess[hash(I')]
    then
      continue with for-loop;
    end–if
  end–macro
```

*Figure 3.11.* Checking for exclusion (early pruning).

```
UPDATE-STATE-DATA-JANUS
  macro
    if I' ∉ next_states or newcost < minguess[hash(I')]
    then
      minguess[hash(I')] := newcost;
      π[hash(I')] := current ordering;
      upper_bound := update_upper_bound();
      if I' ∉ next_states then
        next_states[hash(I')] := I';
        lower_bound[hash(I')] :=
        compute_lower_bound(I');
        if lb_combined_I' ≥ upper_bound then
          exclude I';
        end–if
      end–if
    end–if
  end–macro
```

*Figure 3.12.* Updating the state data in algorithm JANUS.

UPDATE-STATE-DATA-JANUS in Figure 3.12. This saves the cost of unnecessarily repeated computations.

#### 3.1.3.2 Partial BDD Reconstruction

The BDD corresponding to a state is kept in memory only until the next state is considered since otherwise the memory requirement would be much too large. Every BDD for the next state processed (i.e. expanded to its successors) must be reconstructed by variable shifts.

Reconsidering the exact minimization algorithm described in Section 3.1.1, we observe: at step $k$, a BDD $F$ is appropriate to represent a state $I$ with $I = |k-1|$ iff $F$ has a variable ordering $\pi \in \Pi(I)$.

In the approach of [DDG00], the variable ordering used for reconstruction of a BDD for a state $I$ is that of the BDD for $I$ in the previous step $k-1$. This variable ordering $\pi$ respects the above condition $\pi \in \Pi(I)$. The BDD is reconstructed by a series of $n$ upward variable shifts: from left to right, the variables $\pi(1), \ldots, \pi(n)$ are shifted to levels $1, \ldots, n$.

Now suppose, this sequence of variable shifts is reduced to only shifting, from left to right, the variables $\pi(1), \ldots, \pi(|I|)$ to levels $1, \ldots, |I|$. This yields a variable ordering $\pi_I$ which also respects the condition $\pi_I \in \Pi(I)$. The advantage is that much less shift operations are needed. A problem is the higher risk of "BDD explosions". Since only the upper parts of the partially reconstructed BDD and the old BDD for state $I$ coincide, the node number in the lower part of the partially reconstructed BDD can "blow up" which would result in a slow down of run time and an increase of memory requirement. In the approach JANUS, this problem is addressed straightforwardly: whenever 0.7 times the size of the partially reconstructed BDD exceeds the size of the old BDD, the method returns to the old variable ordering $\pi$ which was used to represent state $I$ in the previous iteration. Note that this technique transfers directly to the case of bottom-up minimization.

Additionally, before reconstruction, the algorithm examines the BDDs in a cache holding 10 BDDs and computes the BDD with the smallest number of variable exchanges to set a required ordering. This is done in CPU time much less than the variable exchanges in fact would require. The idea of a BDD cache was introduced by [GD00] and is used here in exact BDD minimization for the first time.

The approach JANUS significantly gains efficiency by using these techniques of partial reconstruction.

### 3.1.4 Experimental Results

All experimental results have been carried out on a system with an Athlon processor running at 1.4 GigaHz using an upper memory limit of 300 MByte and a run time limit of 20,000 CPU seconds. The algorithm presented in the previous sections is called JANUS $\uparrow$ if minimization progresses bottom-up using $l\_b_\omega$ and JANUS $\downarrow$ if minimization progresses top-down using the lower bound $l\_b_\alpha$. Both approaches use the "early pruning" techniques of Section 3.1.2.2. The implementation of the algorithm JANUS is based on the implementation of the algorithm in [DDG00], called FizZ. Both algorithms have been integrated in the CUDD package [Som02] which also contains an algorithm for exact BDD minimization [JKS93]. By this it is guaranteed that all algorithms run in the same system environment.

In a first series of experiments both algorithms were applied to the set of benchmark circuits from LGSynth93 [Col93]. The results are given in Table 3.1. In the first column the name of the function is given. Column *in* (*out*) gives the number of inputs (outputs) of a function. Column *opt* shows the number of BDD nodes needed for the minimal representation. In columns *time* and *space* the run time in CPU seconds and the space requirement in MByte for JUNON and the approach JANUS $\uparrow$ as well as for the approach FizZ and the approach JANUS $\downarrow$ are given. The functions are ordered by ascending run times of the algorithm JANUS $\downarrow$.

As the results show, the algorithm JUNON has much longer run times[5] than the bottom-up approach JANUS $\uparrow$. It can be seen that JANUS $\uparrow$ often accelerates run time by a factor of up to two orders of magnitude (see e.g. *sct, pcle, tcon*) in comparison to JUNON. However, JANUS $\uparrow$ has longer run times than the top-down approaches in most cases. Note that the order of functions by ascending run times of JANUS $\uparrow$ in some cases is different from that order for JANUS $\downarrow$ (which was chosen for Table 3.1). Moreover, there are cases, in which JANUS $\uparrow$ is significantly faster than the top-down approach FizZ, see e.g. *cm150a, mux*. For these cases the run time even comes very near to that of JANUS $\downarrow$.

In a second series of experiments, we give more insight in the influence of the various optimization techniques of JANUS. For this purpose, the reduction in run time in comparison to FizZ has been determined, as it has been gained when applying each technique of JANUS $\downarrow$ separately. These techniques have been integrated individually into the implementation of the algorithm FizZ. The results are given in Table 3.2. The first four columns coincide with the first four columns of Table 3.1, giving the function name, the number of inputs and outputs and the number of BDD nodes needed for the minimal representation.

Column *impr. lb* states the run times obtained by applying the improved lower bound technique described in Section 3.1.3.1: here the algorithm switches between two different routines for the lower bound computation, depending on what routine is more promising in terms of expected run time.

Column *cost. lb* gives the run times achieved by individually applying the $l\_b_{cost}$ lower bound described in Section 3.1.2.2 whereas Column *comb.* gives the run times obtained by use of the lower bound $l\_b_{combined}$

---

[5]In all cases where a memory requirement, but no run time for algorithm JUNON is given, the time limit (also) has been exceeded. In the cases where no value is given at all, the system even failed to allocate the initial amount of memory required by the algorithm.

*Table 3.1.* Comparison of JUNON, FizZ and JANUS

| | | | | bottom-up | | | | top-down | | | |
|---|---|---|---|---|---|---|---|---|---|---|---|
| name | in | out | opt | JUNON | | JANUS ↑ | | FizZ | | JANUS ↓ | |
| | | | | time | space | time | space | time | space | time | space |
| parity | 16 | 1 | 17 | <0.01s | 2M | 0.01s | <1M | <0.01s | <1M | 0.03s | <1M |
| cmb | 16 | 4 | 28 | 0.02s | 2M | 0.04s | <1M | 0.01s | <1M | 0.05s | <1M |
| t481 | 16 | 1 | 21 | 0.16s | 2M | 0.15s | <1M | 0.16s | <1M | 0.13s | <1M |
| tcon | 17 | 16 | 25 | 635s | 4M | 6.61s | 4M | 0.52s | <1M | 0.28s | <1M |
| pm1 | 16 | 13 | 40 | 1.26s | 2M | 0.64s | <1M | 0.55s | <1M | 0.34s | <1M |
| cm163a | 16 | 5 | 26 | 5.23s | 2M | 1.87s | <1M | 1.17s | <1M | 0.78s | <1M |
| cordic | 23 | 2 | 42 | 11.2s | 258M | 5.57s | <1M | 3.05s | <1M | 1.82s | <1M |
| pcle | 19 | 9 | 42 | 7584s | 15M | 60.9s | 8M | 9.02s | 2M | 5.18s | 3M |
| s208.1 | 18 | 9 | 41 | 2116s | 8M | 42.2s | 4M | 8.44s | <1M | 5.62s | 2M |
| sct | 19 | 15 | 48 | 8453s | 15M | 53.3s | 8M | 8.62s | 2M | 5.97s | 3M |
| s298 | 17 | 20 | 74 | 934s | 4M | 25.6s | 4M | 13.46s | 2M | 9.06s | 3M |
| i1 | 25 | 16 | 36 | – | – | 74.1s | 19M | 29.4s | 9M | 18.77s | 10M |
| vda | 17 | 39 | 478 | 822s | 5M | 99.9s | 5M | 65.4s | 3M | 34.4s | 3M |
| cc | 21 | 20 | 46 | – | 62M | 412s | 40M | 117s | 32M | 84.9s | 34M |
| mux | 21 | 1 | 33 | – | 62M | 348s | 35M | 610s | 32M | 311s | 35M |
| cm150a | 21 | 1 | 33 | – | 62M | 348s | 35M | 610s | 32M | 311s | 34M |
| s400 | 24 | 27 | 119 | – | 537M | 3322s | 263M | 802s | 67M | 456s | 71M |
| s382 | 24 | 27 | 119 | – | 537M | 3318s | 263M | 802s | 67M | 461s | 71M |
| lal | 26 | 19 | 67 | – | – | 3818s | 256M | 677s | 64M | 504s | 75M |
| s444 | 24 | 27 | 119 | – | 537M | 3331s | 263M | 779s | 67M | 508s | 78M |
| ttt2 | 24 | 21 | 107 | – | 537M | 5281s | 277M | 950s | 74M | 578s | 78M |
| s526 | 24 | 27 | 113 | – | 537M | 6017s | 284M | 1196s | 93M | 924s | 105M |
| s349 | 24 | 26 | 104 | – | 537M | 2921s | 229M | 1447s | 99M | 950s | 105M |
| s344 | 24 | 26 | 104 | – | 537M | 2923s | 229M | 1446s | 99M | 950s | 105M |
| s820 | 23 | 24 | 220 | – | 258M | 2467s | 76M | 2034s | 53M | 1235s | 56M |
| s832 | 23 | 24 | 220 | – | 258M | 2541s | 69M | 2076s | 53M | 1288s | 56M |
| cps | 24 | 102 | 971 | – | 537M | 9130s | 85M | 4396s | 48M | 2751s | 58M |
| comp | 32 | 3 | 95 | – | – | 8684s | 146M | 5606s | 99M | 3900s | 125M |

described in the same section. Column *early* states the run times that result from applying both "early pruning" techniques of Section 3.1.2.2, i.e. $l\_b_{cost}$ and $l\_b_{combined}$.

*Table 3.2.* Techniques of JANUS↓

| name | impr. lb | cost lb | comb. | early | shifts | cache | reconst. |
|---|---|---|---|---|---|---|---|
| parity | <0.01s | <0.01s | <0.01s | <0.01s | <0.01s | 0.28s | 0.22s |
| cmb | 0.02s | 0.02s | 0.02s | 0.01s | 0.02s | 0.33s | 0.29s |
| t481 | 0.14s | 0.15s | 0.14s | 0.13s | 0.14s | 0.45s | 0.47s |
| tcon | 0.48s | 0.46s | 0.51s | 0.47s | 0.40s | 0.82s | 0.71s |
| pm1 | 0.54s | 0.52s | 0.51s | 0.52s | 0.36s | 0.92s | 0.69s |
| cm163a | 1.16s | 1.07s | 1.10s | 1.05s | 0.78s | 1.47s | 0.99s |
| cordic | 2.77s | 3.00s | 2.65s | 2.64s | 2.65s | 3.07s | 2.71s |
| pcle | 8.56s | 8.03s | 8.53s | 7.65s | 6.85s | 8.77s | 6.92s |
| s208.1 | 8.00s | 7.95s | 7.67s | 7.31s | 7.15s | 8.36s | 7.09s |
| sct | 8.04s | 8.21s | 7.95s | 7.77s | 6.99s | 8.71s | 7.35s |
| s298 | 13.13s | 13.24s | 13.05s | 12.95s | 10.38s | 12.55s | 10.11s |
| i1 | 27.9s | 27.2s | 27.3s | 26.0s | 24.0s | 28.6s | 22.8s |
| vda | 53.5s | 65.0s | 55.9s | 55.9s | 54.3s | 62.1s | 51.0s |
| cc | 116s | 111s | 115s | 111s | 91s | 111s | 92s |
| mux | 602s | 521s | 564s | 478s | 487s | 560s | 429s |
| cm150a | 602s | 522s | 565s | 478s | 489s | 560s | 432s |
| s400 | 726s | 775s | 725s | 710s | 606s | 791s | 585s |
| s382 | 726s | 774s | 726s | 710s | 605s | 793s | 585s |
| lal | 640s | 647s | 609s | 595s | 652s | 630s | 530s |
| s444 | 714s | 747s | 705s | 687s | 707s | 752s | 635s |
| ttt2 | 896s | 903s | 881s | 848s | 739s | 903s | 706s |
| s526 | 1138s | 1173s | 1124s | 1110s | 1047s | 1121s | 991s |
| s349 | 1325s | 1407s | 1342s | 1311s | 1322s | 1341s | 1226s |
| s344 | 1325s | 1408s | 1342s | 1311s | 1316s | 1339s | 1227s |
| s820 | 1863s | 2042s | 1876s | 1876s | 1699s | 1920s | 1538s |
| s832 | 1873s | 2076s | 1905s | 1902s | 1729s | 1970s | 1590s |
| cps | 3999s | 4395s | 3842s | 3833s | 4215s | 4034s | 3700s |
| comp | 5135s | 5868s | 4978s | 5190s | 6509s | 4274s | 4929s |
| avg. gain (%) | 8.0 | 0.8 | 9.6 | 10.2 | 5.7 | 10.4 | 18.5 |
| max. gain (%) | 18.2 | 14.4 | 14.5 | 18.8 | 34.5 | 23.8 | 29.2 |

In column *shifts* the obtained run times using a much shorter sequence of variable shifts for BDD reconstruction as described in Section 3.1.3.2 are given. The next column *cache* states the run times using the BDD cache as explained in the same section. Finally, in column *reconst.*, the results of applying these two techniques (shorter shift sequence/cache) together are presented.

In the last two rows of Table 3.2 the obtained run times with the obtained average and maximal gain for each technique are summarized. The "early pruning" techniques of Section 3.1.2.2 yield a gain of up to 18.8% and a reduction in run time of 10.2% on average whereas the techniques of partial reconstruction of Section 3.1.3.2 result in a gain of up to 29.2%. On average, the reduction in run time is 18.5%.

Consequently, the top-down approach JANUS$\downarrow$ is faster than FizZ, especially for larger examples achieving a reduction in run time by up to 49% (see e.g. *mux*). On average, the reduction in run time is 35.4%. The results show that the novel threefold lower bound technique together with efficient implementation techniques is a very robust improvement that significantly outperforms the original algorithm FizZ.

## 3.2 $A^*$-based Optimization

The latest development in the field of exact BDD optimization is a shift to a new paradigm, the *$A^*$-algorithm.* This is a search technique frequently used in AI.

In this section, first the theoretical background of the $A^*$-algorithm is given. Next, an exact BDD minimization algorithm is presented which is based on this paradigm. In terms of run time, it is superior to all B&B-based approaches described in the previous section. Moreover, ordered best-first search, i.e. the $A^*$-algorithm, can be *combined* with a classical B&B algorithm. This allows to avoid unnecessary computations and to save memory.

### 3.2.1 State Space Search by $A^*$-Algorithm

In a nutshell, $A^*$ operates on a state space. Large parts of it are pruned by a best-first strategy expanding only the most promising states.

A state space problem consists of determining a sequence of operators

$$\xrightarrow{O_1}, \xrightarrow{O_2}, \ldots, \xrightarrow{O_n}$$

that, when applied to the initial state, yields a goal state. An important method to *guide* the search on a state space is *heuristic search.* With every state $q$ a quantity $h(q)$ is associated which estimates the cost of the cheapest path from $q$ to a goal state. This allows us to search in the direction of the goal states. The *$A^*$-algorithm* as introduced by [HNR68] bases the choice of the next state to expand on two criteria: the cost of the cheapest known path from the initial state up to state $q$, denoted $g(q)$, and the estimate $h(q)$ which for $A^*$ has to be a lower bound on the cost of an optimal path from $q$ to a goal state (this minimal cost is denoted $h^*(q)$). $A^*$ maintains a prioritized queue OPEN which is ordered with respect to increasing values $\varphi(q) = g(q) + h(q)$, thus *combining* the two criteria

a) cheapest path to $q$ known so far and

b) expected cost of the remaining part of the path from $q$ to a goal state.

The function $\varphi$ is called the *evaluation function.* In the beginning, this queue only contains the initial state. In each step, a state $q$ with a *minimal* $\varphi$-value is expanded and dequeued. During expansion, the successor states of $q$ are generated and inserted into the queue according to their $\varphi$-values. For this, the values $g$ and $h$ of the successor states are computed dynamically. For a transition $q \longrightarrow q'$ let $c(q, q')$ denote the cost of the transition. Then $q'$ is associated with its cost $g(q') = g(q) + c(q, q')$, i.e. $g$ accumulates transition costs. In this, for a state $q$, $g(q)$ is computed as the sum of the cost $c(r, r')$ of all transitions $r \longrightarrow r'$ occurring on the cheapest known path to $q$.

A successor state $q'$ might be generated a second time if $q'$ has more than one predecessor state. If a cheaper path from the initial state to $q'$ is found in this case, $g(q')$ is updated. These updates of the "$g$-part" of $\varphi$ to the costs of a newly found cheaper path to $q$ continously compensate for the fact that the character of the "$h$-part" is only estimative. The cheapest known path to $q'$ is denoted $p(q')$ and is also updated respectively. The algorithm terminates if the next state to expand is a goal state $t$. The estimate $h(t) = h^*(t)$ must be zero. In this case, the path found up to $t$, i.e. $p(t)$, is of minimal cost $C^*$ which is also expressed with $\varphi^*(t) = g(t) = g^*(t) = C^*$. This minimum cost path $p(t) = p^*(t)$ is reported as solution. A deep analysis of this strategy shows that $A^*$ always terminates and finds such a minimum cost path to a goal state if $h$ is *admissible*, i.e. $h$ never overestimates the true cost [HNR68, Pea84]. Formally, a heuristic function $h$ is admissible iff for all states $q$, $h(q)$ is a lower bound on $h^*(q)$.

In AI, a classical example for heuristic state space search is the $((n \times n) - 1)$-puzzle: $((n \times n) - 1)$ tiles, each labeled with a number in $\{1, 2, \ldots, n^2 - 1\}$, are placed on a square board with $(n \times n)$ fields. Starting from an arbitrary distribution, the aim is to reach the ordered goal distribution, see an example for $n = 3$ in Figure 3.13. A valid move from one distribution to another is done by shifting a tile upwards, downwards, left or right to the one free field of the board. For that, the tile must be positioned on a field adjacent to this free field (see the sample move in Figure 3.14).

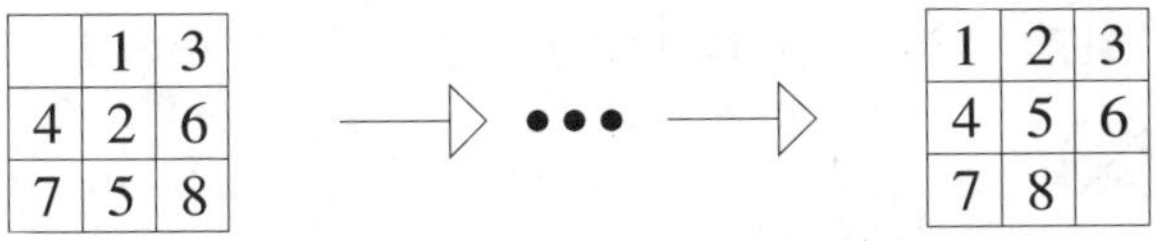


*Figure 3.13.* Solving a $((3 \times 3) - 1)$-puzzle.

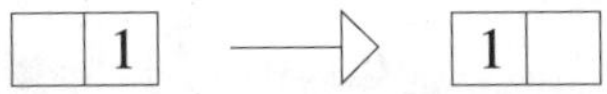


*Figure 3.14.* A valid move in the puzzle.

The considered state space is the set of all distributions, each encoded as one state $q$. The cost of a transition from one distribution to another, i.e. the cost of one such move, is considered to be constant, e.g. $c(q, q') = 1$ for all moves $q \longrightarrow q'$. Then $g(q)$ maintains the smallest known number of moves needed to reach a particular distribution $q$ from the initial distribution.

There are several possible choices for an appropriate heuristic function $h$· in the following, two reasonable estimates for use in an $A^*$-algorithm to solve the $((n \times n) - 1)$-puzzle are given:

1) $h_1(q)$ = number of tiles on "wrong" positions in $q$, and

2) $h_2(q)$ = sum of the horizontal and vertical distances of the tiles in $q$ to their positions in the goal distribution.

Clearly, both heuristic functions are lower bounds on $h^*(q)$, i.e. they are admissible.

In this example it has been illustrated, how the formal framework of the $A^*$-algorithm problem can be applied to many problem spaces, i.e. an example for the choice of the respective quantities has been given. Next, a more formal example is given to illustrate the flow of the algorithm. For this purpose, Figure 3.15 illustrates, how a state space is traversed with the $A^*$-algorithm: first, an initial state $a$ is expanded and yields three successor states $b$, $c$, and $d$. The cost of a transition is annotated at the respective edge, i.e. the transition $a \longrightarrow c$ is of cost 2. At a state $q$, the values $g(q)$ and $h(q)$ are annotated in the form "$g(q)|h(q)$". In the example, $g(c) = 2$, $h(c) = 4$ and thus $\varphi(c) = g(c) + h(c) = 6$. After expansion of the initial state $a$, $a$ is removed from the queue Open which holds $d$, $b$, and $c$ in the order of increasing $\varphi$-values. The state with minimal $\varphi$-value is chosen as the next to expand: this is $d$ with $\varphi(d) = 3$. Expansion yields states $e$ and $f$ and $d$ is removed from the queue Open. This process continues in the next steps depicted in Figure 3.15. Note the situation in the step before last where states $h$

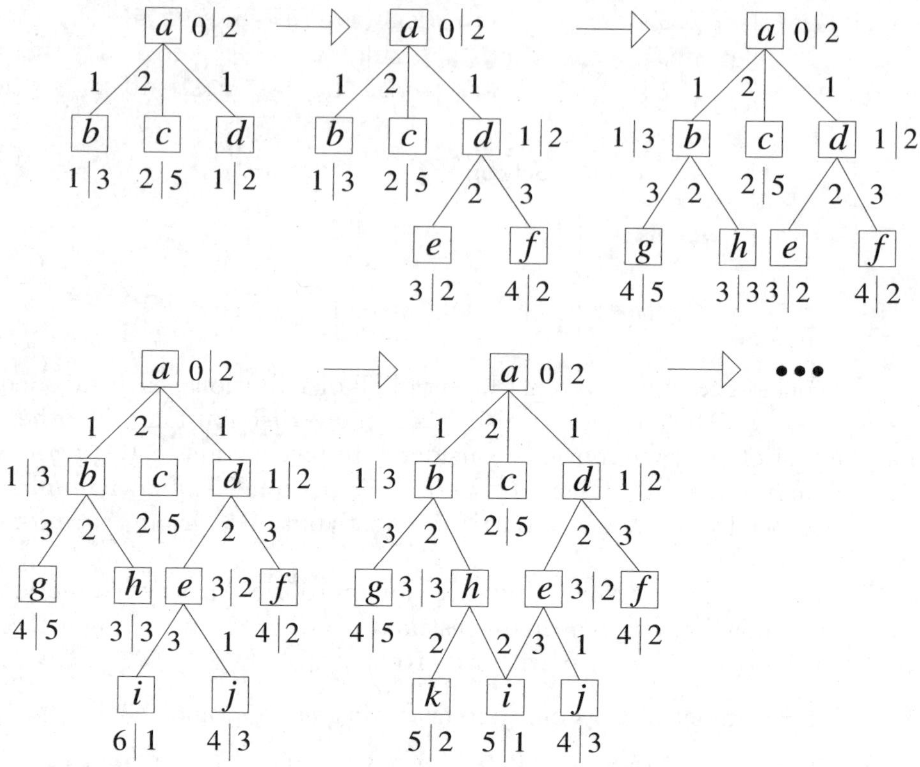


*Figure 3.15.* State Space Search by $A^*$.

and $f$ both have minimal total cost $\varphi(h) = \varphi(f) = 6$. In this case of a tie between several states some *tie-breaking rule* must exist to choose one of the candidates. In the example, state $h$ is chosen as the next state for expansion. The creation of a state can result from more than one expansion: in the example, state $i$ has a best known path with a cost of 6 in the step before last. In the next step, state $h$ is expanded, yielding state $i$, introducing a new path to $i$ from the initial state via $h$. This path is only of cost 5, and thus both $g(i)$ and $p(i)$ are updated accordingly. This process terminates as soon as a goal state is found.

Note that *not all* quantities mentioned in the above description of the $A^*$-algorithm are well-defined functions: yet the quantities $\varphi$, $g$ and $p$ are associated with a state $q$, but their values continously change during the algorithm run (i.e., the algorithm may *update* the information associated with states $q$). Hence $\varphi$, $g$, and $p$ should be thought of as temporary mappings of states to values (in $\mathbb{N}$ or sometimes in $\mathbb{R}$) which, together with other informations, define the state, the computation of the $A^*$-algorithm is currently in. It is possible to formalize the process of an $A^*$-computation as a sequence of temporal state vectors (sometimes called interpretations), giving the mappings $\varphi$, $g$ and $p$ which are valid at a certain stage of computation. However, all results presented in this book can be proven without such a formal framework. Throughout the book, these quantities are thus denoted as function values. The results presented are *inequalities* over these values. They express conditions which *hold for all times* (during an $A^*$-computation), in the sense of a temporal tautology. Hence the outlined formalization with temporal interpretations is not necessary. Whenever it is noteworthy, it will be mentioned that the quantities $\varphi$, $g$, and $p$ are subject to changes during the algorithm run.

However, the quantities $h$ and $c$ (heuristic *function* $h$ and transition cost *function* $c$) indeed must be well-defined functions in every $A^*$-algorithm. In our context of exact BDD minimization, this will be verified in later sections.

A state can be seen as an instance of a subproblem: instead of having to find an optimal path from the initial state to a goal state, the subject is to look for a path starting at the considered state. Unlike other search strategies, $A^*$ has a maximum capability of avoiding the consideration of subproblem instances, see [DP87]. To this end, $A^*$ needs to store all active subproblems in OPEN, making it possible to delay the processing of subproblems as far as possible while still guaranteeing minimality of the solution. With this strategy, many subproblems are delayed "forever", they are not considered at all during run time. In this sense, the strategy to always choose the most promising subproblem first (based on the values of the evaluation function) is much more than just a greedy algorithm using some sort of cost function. A drawback of $A^*$ however is the higher memory requirement which is exponential in general: $A^*$ needs to store the active (i.e. "open") subproblems whose number can get large during the algorithm run.

Although the problem of high memory requirement is known in the context of exact BDD minimization (all algorithms suggested so far show an exponential space demand), this clearly is a problem which must be addressed to preserve practicality. Therefore later an effective remedy for this issue is provided, saving more than 60% of the required memory of exact BDD minimization. Moreover, this technique yields an advanced implementation of the generic $A^*$-algorithm. This optimization can be transferred to other applications.

$A^*$ is known to be *optimal* in the class of heuristic search algorithms that use the same evaluation function and that are guaranteed to find an optimal solution: larger parts of state spaces are pruned by $A^*$ than for any other such algorithm [DP87]. More exactly, $A^*$ is known to minimize the number of distinct expanded states. In case of a *monotone* heuristic function $h$, every expanded state is "opened" exactly once. If this assumption of monotonicity is not met, $A^*$ may *reopen* a state several times. In the worst-case this may result in a number of reinsertions of a state into OPEN which grows exponentially in $N$ where $N$ is the number of distinct expanded states. This problem was pointed out and addressed in a new improved algorithm called $B$ by [Mar77]. This algorithm still requires $O(N^2)$ expansions in the worst-case. 1984, Mérõ presented the most recent improvement of algorithm $B$, an algorithm called $B'$ [Mo84] which however has a worst-case complexity of the same order, $O(N^2)$, as algorithm $B$.

Hence, to ensure efficiency of $A^*$-based approaches, the property of *monotonicity* of $h$ is strongly desired. It guarantees a worst-case behavior which is only linear in $N$. In the next sections a formal definition of monotonicity will be given and this property is proven for the heuristic function used in the $A^*$-approach.

The above criterion for optimality, i.e. the number of state expansions, establishes a simplified, abstract view which is justified by the fact that state expansions normally dominate the run time of the algorithm. This is also true in the context of an $A^*$-based algorithm for exact BDD minimization, as every state expansion involves the following two time consuming operations: first, a BDD representing the considered state must be reconstructed and second, a BDD variable must be shifted. However, despite the outlined optimality of $A^*$, of course it is still possible to further improve the performance of $A^*$. In fact in a later section it can be seen, how combination of $A^*$ with B&B and other techniques reduce both memory requirement and run time of the algorithm.

### 3.2.2 Best-First Search Algorithm

Suppose the subject is to minimize a BDD representing a Boolean multi-output function $f: \mathbf{B}^n \rightarrow \mathbf{B}^m$. In the latest approach, the $A^*$-algorithm is used to search for the optimal variable ordering of a BDD. Instead of searching the whole set of variable orderings (which contains $n!$ orderings), the algorithm operates on a state space. This space is $2^{X_n}$, the set of variable sets which is a space of a size growing much slower than $n!$. A state of this space is a set of variables $q \subseteq X_n$. This corresponds to the use of variable sets in B&B methods as described in Section 3.1. However, instead of an explicit enumeration of all possible variable sets and the corresponding variable orderings, an exploration of only the necessary parts of the state space is performed based on expansion of the most promising states.

Both in the B&B-based approaches of Section 3.1 and in the recent best-first search approach, a set $q \subseteq X_n$ represents all orderings of a BDD in which the first $|q|$ positions (or last $|q|$ positions, respectively) constitute $q$. Lemma 3.1 motivates why it is sufficient to consider variable sets instead of considering every single represented ordering.

#### 3.2.2.1 Optimal Ordering by Path Cost Minimization

In the following it is explained how the problem of finding an optimal variable ordering can be expressed as the problem of finding a minimum cost path from an initial state to a goal state in this state space. Doing so, one has to consider paths from the initial state via other states to the goal state. As there are many paths via one state (their number is the product of ingoing and outgoing paths) one might expect this to introduce efficiency problems. Later sections make clear that this is not the case: by the properties of $A^*$, the run time is dominated by state expansions only. Hence, in fact states are explored rather than the paths through them, making it possible to benefit from the smarter encoding of the search problem as outlined in the previous section.

Sets of variables $q$ successively growing from $\emptyset$ to $X_n$ are considered: $q$ is extended at each transition by a variable $x_i \in X_n \setminus q$, i.e. $q \xrightarrow{x_i} q \cup \{x_i\}$. The algorithm starts in the initial state $\emptyset$ and progresses until the goal state $X_n$ is reached. As is described before in Section 3.2.1, $A^*$ finds a path $p^*(X_n)$ from $\emptyset$ to $X_n$ with minimal cost. This cost is the accumulated transition cost for the transitions

$$\emptyset \xrightarrow{x_{i_1}} \{x_{i_1}\} \xrightarrow{x_{i_2}} \cdots \xrightarrow{x_{i_n}} X_n.$$

The sequence of variables occurring on path $p^*(X_n)$ defines a variable ordering.

The basic idea of the $A^*$-based approach is the following: we want the above ordering annotated along the minimum cost path to be optimal, i.e. we want the minimal cost for the goal state $X_n$ (i.e. $\varphi^*(X_n)$) to be the number of nodes in the BDD with this annotated order $x_{i_1} < x_{i_2} < \cdots < x_{i_n}$. Then the sequence of variables in $p^*(X_n)$, describing a minimum cost path from $\emptyset$ to $X_n$, is an optimal variable ordering which yields the minimal BDD size.

#### 3.2.2.2 Cost Function

For this purpose, an appropriate cost function is chosen. In Section 3.2.2.1 it was explained that the variable sequences annotated along paths from the initial state to the goal state can be interpreted as variable orderings in BDDs. With that notion, the annotations of paths to non-goal states have the meaning of *prefixes* of variable orderings, i.e. a path of length $k$ defines the positions of the first $k$ variables in a variable ordering. The key idea now is to define the cost function such that the number of nodes in the first $k$ levels of a BDD is taken as the cost of this path.

In detail: assume a state $q$ always is associated with a BDD whose first $|q|$ variables are ordered according to the first $|q|$ positions of the variable ordering which is annotated at the transitions of the currently considered (i.e., cheapest known) path to $q$. With that, this BDD respects a variable ordering $\pi$ with $\pi \in \Pi(q)$. As cost of the transition $q \xrightarrow{x_i} q \cup \{x_i\}$ the nodes labeled with variable $x_i$ in the BDD which corresponds to the successor state $q \cup \{x_i\}$ are counted. (The variable $x_i$ resides at the $(|q|+1)$-th level of this BDD.) In $g$ the transition costs along the newly found path to the successor are accumulated. Further, for the initial state we set $g(\emptyset) = 0$. Now, by this inductive definition of $g$, the accumulated cost $g(q)$ for each state $q \subseteq X_n$ associated with a BDD $F$ as described above is the number of BDD nodes labeled with a variable in $q$. The cost for the goal state $X_n$ now is (as intended) the size of the BDD associated with it. This BDD has the variable ordering annotated along the minimum cost path by construction. Thus, this variable ordering must be in fact optimal. Figure 3.16 demonstrates the idea of the construction for a transition $q := \{x_1, x_2\} \xrightarrow{x_4} q \cup \{x_4\}$ with BDDs associated with $q$ and the successor state. Nodes with variables belonging to the state represented by the BDD are shaded.

With that it is already shown that the problem of finding an optimal variable ordering can be interpreted and solved as a problem of finding a minimum cost path.

Typically, in AI the $A^*$-algorithm is applied to problem domains where

- the transition cost function is much simpler than the one described here, e.g. often the transition cost is *constant*,
- the set of goal states is pre-defined by some terminating condition in advance, i.e. we "know, what we are looking for".

E.g., this holds for the $((n \times n) - 1)$-puzzle[6] described in Section 3.2.1: every move of a tile is considered to have a constant cost of 1, and there is exactly one pre-defined terminating order of tiles.

Both does not hold when applying the generic $A^*$-algorithm to exact BDD minimization: the transition cost essentially depends on the two

[6]The $((n \times n) - 1)$-puzzle is known to be NP-complete, e.g. see [Ric88].

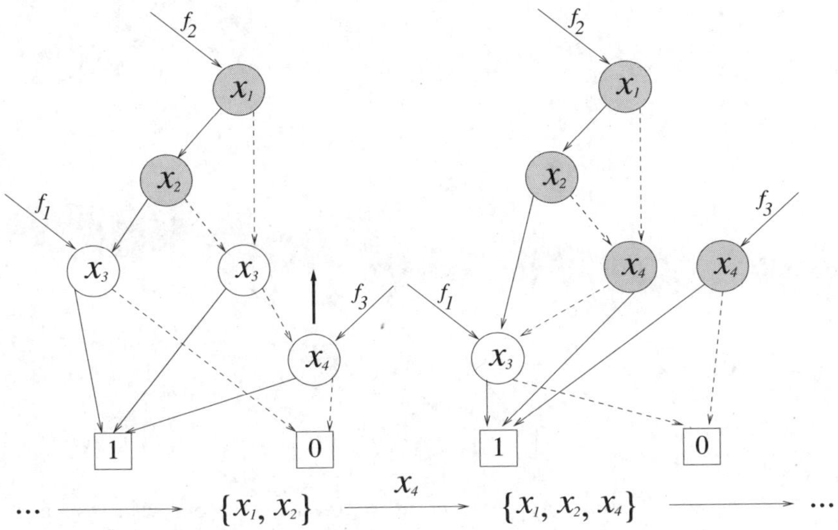


*Figure 3.16.* BDDs for a transition $q := \{x_1, x_2\} \xrightarrow{x_4} q \cup \{x_4\}$.

states of the considered transition and there does not exist any predefined condition defining goal states, i.e. we do not know in advance, towards which variable ordering the search should proceed.

In this, the idea to apply $A^*$ to the problem of determining an optimal variable ordering of BDDs is not an obvious one. Moreover, when compared to other problem domains, it is conceptually harder to suggest good heuristics to guide the search through the state space since we do not know in advance, what the result (i.e., the optimal ordering) will look like.

To summarize: $A^*$ prunes large parts of the state space by a strategy to always choose the "best", i.e. most promising state first for the next state expansion. The decision, whether a state $q$ is "promising" or not, is made on the basis of the evaluation function $\varphi(q) = g(q) + h(q)$.

It consists of a part keeping track of the currently known minimal cost for a path to $q$, $g(q)$, and a heuristic estimate $h(q)$ of (i.e. a lower bound on) the cost of the (not yet known) path from $q$ to the goal state $X_n$. It is computed for every state in a set of "open" states OPEN. The next state to expand is a state $q_{\text{next}} \in$ OPEN with

$$\varphi(q_{\text{next}}) = \min\{\varphi(q) \mid q \in \text{OPEN}\}. \tag{3.3}$$

The BDD associated with $q_{\text{next}}$ is reconstructed according to the construction described before.

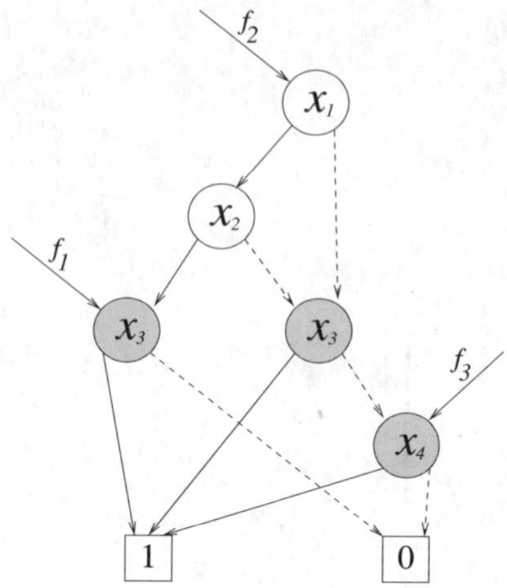


*Figure 3.17.* A shared diagram. Gray nodes represent cofactors with respect to $x_1$ and $x_2$.

#### 3.2.2.3 Heuristic Function

Next the heuristic function used in the $A^*$-based approach is given as $h_{f,n}: 2^{X_n} \rightarrow \mathbb{N}$

$$h_{f,n}(q) = \max\left(\left|\mathrm{cof}(f,q)\right|, n - |q|\right). \tag{3.4}$$

The Boolean function $f$ and its arity $n$ are always given from the context. Hence in accordance with the standard notations in the $A^*$-related literature, the heuristic function from now on will be denoted $h$ throughout the book, omitting the subscripts $f$ and $n$.

This heuristic function is essentially the one proposed in Section 3.1, a lower bound adapted from VLSI design. However, unlike in Section 3.1, the lower bound is expressed in Equation (3.4) as a *function* of $f$ and the state $q$ only. With that the bound becomes *independent* of the graph structure of a BDD representing state $q$. This can be used for a more efficient implementation, avoiding unnecessary computations. This is explained briefly after introducing the lower bound. For a state $q$, the lower bound counts the number of cofactors with respect to the variables in $q$ (see Figure 3.17). The reader may recall that, more formally spoken, the set $\mathrm{cof}(f,q)$ is the set of all distinct (non-constant single-output) cofactors of $f$ ($f$ is interpreted as a family of $n$-ary single-output functions) with respect to all variables in $q$. Also note that this is not a multiset, hence functionally equivalent cofactors are eliminated and thus do not contribute to $|\mathrm{cof}(f,q)|$ (see Section 2.3).

Let $\pi \in \Pi(q)$ and $F := \mathrm{BDD}(f,\pi) = (\pi, \ldots, O)$. As can already be seen with Figure 3.17, nodes representing the considered cofactors

$\mathrm{cof}(f,q)$ essentially are nodes in the lower part referenced from nodes of the upper part or they are output nodes in the lower part. More formally, these are the nodes $\mathrm{ref}(F,|q|) \cup O_n^{|q|+1}$ (where $O_n^{|q|+1} = O \cap F_n^{|q|+1}$, see Chapter 2). This intuitive correspondence of the lower bound introduced in Section 3.1 to the function $h(q)$ is subject to the following formal proof.

DEFINITION 3.9 *Let*

$\mathcal{K}$: $\{F \mid F \text{ is a BDD over } X_n\} \times \mathbb{N} \to 2^{\{v\in V|(\ldots,(V,E),\ldots) \text{ is a BDD}\}}$;

$$\mathcal{K}(F,k) = \mathrm{ref}(F,k) \cup O_n^{k+1}$$

LEMMA 3.10 *Let* $f\colon \mathbf{B}^n \to \mathbf{B}^m$, $q \subseteq X_n$, $\pi \in \Pi(q)$ *and* $F := \mathrm{BDD}(f,\pi)$. *Then we have*

$$\begin{aligned} |\mathrm{ref}(F,|q|)| &= |\mathrm{tcof}(f,q)|, \\ |\mathcal{K}(F,|q|)| &= |\mathrm{cof}(f,q)|. \end{aligned}$$

**Proof.** First let $q = \emptyset$. By definition we have $|\mathcal{K}(F,0)| = |\mathrm{ref}(F,0)| = |O| = |\mathrm{tcof}(f,\emptyset)| = |\mathrm{cof}(f,\emptyset)|$ where $F = (\pi,\ldots,O)$. Now let $q \supset \emptyset$ and let $g$ be an $n$-ary Boolean function. We show[7]

$$\begin{aligned} (\exists_1 v\colon\ v \in \mathrm{ref}(F,|q|) \text{ and } v \text{ represents } g\ ) &\iff g \in \mathrm{tcof}(f,q), &(3.5)\\ (\exists_1 v\colon\ v \in \mathcal{K}(F,|q|) \text{ and } v \text{ represents } g\ ) &\iff g \in \mathrm{cof}(f,q), &(3.6) \end{aligned}$$

then the required existence of a bijection between $\mathrm{ref}(F,|q|)$ and $\mathrm{tcof}(f,q)$, respectively $\mathcal{K}(F,|q|)$ and $\mathrm{cof}(f,q)$, follows.

With $\pi \in \Pi(q)$ we have $q = \{\pi(1), \pi(2), \ldots, \pi(|q|)\}$. (*)

"$\Longrightarrow$":

Let $v$ be the unique node in one of the requirements. By definition, $v$ is referenced directly from a node in levels $1, 2, \ldots, |q|$ or it is an output node in levels $|q|+1, \ldots, n$. Since each component of a BDD is a connected graph, there is at least one path from an output node to $v$. Along this path at most variables in $q$ are tested by Equation (*). For any variable which is not tested, any binding in $\mathbf{B}$ can be assumed without changing the considered path. Now, along the considered path, a variable is fixed to a constant iff it is contained in $q$. But then the function represented by $v$ must be contained in $\mathrm{tcof}(f,q)$, if $v$ is referenced directly from the upper part or otherwise, it must be contained in

[7] The quantifier symbol "$\exists_1$" has the semantic "there exists exactly one ...".

$\mathrm{cof}(f, q)$ by the BDD semantics of Section 2.4.

"$\Longleftarrow$":

Let $g \in \mathrm{tcof}(f, q)$ or $g \in \mathrm{cof}(f, q)$. By Equation $(*)$ there is an $i$ $(1 \leq i \leq n)$ such that

$$g = f_i|_{\pi(1)=b_1, \pi(2)=b_2, \ldots, \pi(|q|)=b_{|q|}}$$

where $b_k \in \mathbf{B}$ $(1 \leq k \leq |q|)$. By the BDD semantics (see Section 2.4), there must be a node $v$ in $F$ representing $g$ and by Theorem 2.17, this node $v$ must be unique in $F$. Moreover, there is a path $p$ from the output node representing $f_i$ to $v$. Along path $p$, at most the variables in levels $1, 2, \ldots, |q|$ are tested, hence

**(a)** $v$ is directly referenced from a node in levels $1, 2, \ldots, |q|$ or it is an output node in levels $|q| + 1, |q| + 2, \ldots, n$.

The function $g$ does not essentially depend on a variable in $q$ by definition. Since $v$ represents $g$,

**(b)** $v$ must reside in levels $|q| + 1, |q| + 2, \ldots, n$ of $F$.

The intermediate results (a) and (b) yield $v \in \mathrm{ref}(F, |q|)$ in the case, $v$ is referenced directly from the upper part and $v \in \mathcal{K}(F, |q|)$ otherwise. □

Lemma 3.10 considers BDDs respecting an ordering that preserves the partition $(q, X_n \setminus q)$ of the input variables. This partition defines the variables for the upper and lower part of the BDD. The lemma considers the set of nodes in two forms of "kernels": a kernel is defined either as a) the set of nodes directly referenced from nodes in the upper part or b) as an extension of this kernel by the lower output nodes. The result of the lemma now is that both kernel sizes are *invariant* with respect to the choice of a variable ordering respecting the given partition of input variables. The constant kernel sizes are given as either the number of a) *true* or b) general cofactors of $f$ with respect to variables in $q$: these sets of cofactors remain unchanged *regardless of the chosen ordering*, as long as the ordering yields the same partition of input variables; and the nodes in the kernel are needed to *represent* these cofactors.

By that, this result is stronger than the one stated in Lemma 3.4, as Lemma 3.4 does not state that the lower bound is constant, i.e. invariant with respect to choice of an ordering preserving the partition of the inputs. Moreover, the lower bound is expressed such that it depends on the graph structure $F$ (which *changes* with different orderings $\pi \in \Pi(q)$) whereas Lemma 3.10 expresses the lower bound depending on $f$ and the partition only.

Next it is explained why $h(q)$ is a lower bound on the cost of the remaining path from $q$ to the goal state: in the above construction to compute the optimal variable ordering by $A^*$, a state $q$ has been associated with a BDD $F$ having a variable ordering

$$\pi \in \Pi(q). \tag{3.7}$$

By construction it suffices to see that $h(q)$ is a lower bound on the minimal node number $\text{min\_cost}(X_n \setminus q, q)$ in the "lower part" of $F$. All cofactors in $\text{cof}(f, q)$ must be represented by nodes in levels $|q|+1, \ldots, n$ by Equation (3.6) in Lemma 3.10 (e.g. the gray nodes in Figure 3.17 represent the cofactors of $f_1$ and $f_2$ with respect to $x_1, x_2$ and $x_3$).

Thus, $\text{cof}(f, q)$ is a lower bound on $\text{min\_cost}(X_n \setminus q, q)$. In every level from $|q|+1$ to $n$ there is at least one node since $f$ is assumed to depend essentially on all input variables. This yields the second lower bound $n - |q|$. Hence, the maximum of both lower bounds must also be a lower bound.

This lower bound is *tight*: the existence of nodes different from the roots of sub-graphs representing the cofactors is not generally guaranteed. This is because these root nodes might coincide with nodes of sub-graphs which are representing other cofactors, e.g. see the gray node labeled $x_4$ in Figure 3.17: it is both a root node for a cofactor and a node of a sub-graph representing a second cofactor. Experimental results showed a large pruning power of this lower bound, resulting in speed-ups of two orders of magnitude compared to trivial bounds. It can be computed effectively with a top-down graph traversal on the BDD, counting the number of direct references from the upper nodes to the nodes in the lower part of the BDD (see Section 3.1.2.1).

The description of the lower bound is finished by briefly explaining its efficient implementation for the $A^*$-based approach. Reconsider Equation (3.4): in contrast to the definition in Section 3.1.2.1 the bound is expressed as a *function* of $f$ and state $q$ only. Thus, it must be *independent* of the BDD $F$ representing the state $q$ in a step of the algorithm. Consequently, if $q$ is encountered a second time during the algorithm run, $h(q)$ does *not* require a re-calculation. In particular this holds even though the graph structure of the BDD $F$ representing $q$ has been changed in the meanwhile. Yet, as for algorithm JANUS presented in Section 3.1, $F$ is also required to respect an ordering $\pi \in \Pi(q)$ in order to ensure correct calculation of the number of direct references of nodes in the upper BDD part (i.e. nodes labeled with a variable in $q$) to nodes in the lower BDD part (labeled with a variable in $X_n \setminus q$). But this condition $\pi \in \Pi(q)$ is always met by construction whenever a state $q$ is encountered a second time (see Equation (3.7)).

#### 3.2.2.4 Exploring Paths during State Space Exploration

In the beginning of Section 3.2.2 it was stated that the algorithm explores states rather than the paths through them. Next reasons are given, why considering paths in fact is done with only minor additional effort in time and space. First, it is not necessary to store all possible paths to a state. Instead it suffices to store only the cheapest path found so far as one state attribute $p(q)$. This quantity is continously updated by the algorithm each time, a cheaper path is found, just as is done with $g(q)$, the cost of the currently cheapest path. Second, by the properties of $A^*$, we have the following result [HNR68, Pea84].

PROPOSITION 1 *Consider an $A^*$-algorithm with a monotone*[8] *heuristic function $h$. Then, if a state $q$ is expanded, a cheapest path to $q$ has already been found, i.e. we have $g(q) = g^*(q)$ and therefore $p(q) = p^*(q)$.*

Now note that a state $q$ has only $|q|$ potential predecessors, e.g. state $q = \{x_1, x_2, x_3\}$ has the possible predecessors

$$\{x_1, x_2\}, \{x_1, x_3\} \text{ and } \{x_2, x_3\}.$$

To determine the cheapest path to $q$ it suffices to examine at most $|q|$ paths via the predecessors of $q$. This is because, by Proposition 1, the paths to a predecessor $p$ must already be optimal at the time of a transition $p \longrightarrow q$. Moreover, some of the $|q|$ predecessors might even not be expanded during the algorithm run, further reducing the number of paths to consider.

The update of $p(q)$ is done based on a simple comparison of the cost of the path via a new predecessor and the old value of $p(q)$. These updates are operations with only minor run time which can be neglected compared to other dominating sources of run time complexity, e.g. the reconstruction of a BDD and the shifting of BDD variables, both needed once for every state expansion.

### 3.2.3 Monotonicity

In this section, the desired *monotonicity* of $h$ in the $A^*$-algorithm outlined above is proven. The results of this section ensure efficiency of the $A^*$-approach to exact BDD minimization. Moreover, in Chapter 4 proofs of soundness and tightness of the best known lower bound for dynamic reordering are given which are directly based on results of this section.

---

[8] The heuristic function $h$ proposed here is in fact monotone which is shown in Section 3.2.3.

First, a formal definition of the important property of monotonicity and a theorem from search theory are given.

DEFINITION 3.11 *Consider an $A^*$-algorithm with the heuristic function $h$ and the transition cost function $c$. The Heuristic function $h$ is said to be* monotone *if*

$$h(q) \leq c(q, q') + h(q') \tag{3.8}$$

*for every transition $q \longrightarrow q'$.*

Note that according to the definition in Section 3.2.1, heuristic functions $h$ must respect the condition

$$h(q) = 0 \text{ for every goal state } q.$$

This is crucial in the context of monotonicity: e.g., without this condition, the expected costs of each state could be overestimated by a (huge) constant $c$ while still keeping Inequality (3.8) valid.

The following important two results can be found in [HNR68, Pea84].

PROPOSITION 2 *Consider an $A^*$-algorithm with a monotone heuristic function $h$. Then $h$ is also an* admissible *heuristic function, i.e. $h(q) \leq h^*(q)$ for all states $q$. Moreover, $q \longrightarrow q'$ always implies $\varphi(q) \leq \varphi(q')$ iff $h$ is monotone.*

THEOREM 3.12 *Consider an $A^*$-algorithm with a monotone heuristic function $h$. Then a state $q$ is expanded* at most once *by $A^*$.*

As explained at the end of Section 3.2.1, this is crucial for the efficiency of $A^*$, as otherwise states can be expanded "exponentially often" in the worst-case, making the use of $A^*$ impractical for larger problems.

For this reason, the desired monotonicity of the heuristic function $h$ proposed in Section 3.2.1 must be verified. Next, after the introduction of a notation, a well-known result following [FS90] is given.

DEFINITION 3.13 *Let $f\colon \mathbf{B}^n \rightarrow \mathbf{B}^m$ be a Boolean multi-output function essentially depending on all variables in $X_n$ and let $x_i \in X_n$. The set of those cofactors of $f$ with respect to variables in $q$ which essentially depend on $x_i$ is denoted with* $\mathrm{dep}(f, q, x_i)$. *Formally, the function* dep *is given as* dep: $\{f \mid f\colon \mathbf{B}^n \rightarrow \mathbf{B}^m\} \times 2^{X_n} \times X_n \rightarrow 2^{\{f \mid f\colon \mathbf{B}^n \rightarrow \mathbf{B}^m\}}$*;*

$$\mathrm{dep}(f, q, x_i) = \{c \in \mathrm{cof}(f, q) \mid c \textit{ essentially depends on } x_i\}.$$

Now a result is stated which is essentially following an analysis in [FS90]:

LEMMA 3.14 *Let $f\colon \mathbf{B}^n \to \mathbf{B}^m$ be a Boolean multi-output function, let $q \subseteq X_n$ and $\pi \in \Pi(q)$ such that $\pi(|q|+1) = x_i$ and let $F := \mathrm{BDD}(f,\pi)$. Then we have:*

> *The number of nodes in the $(|q|+1)$-th level of $F$ is equal to $|\mathrm{dep}(f,q,x_i)|$,*

*or, more formally,*

$$|\mathrm{level}(F,|q|+1)| = |\mathrm{dep}(f,q,x_i)|\,.$$

A more detailed formulation of this result is[9]

$$(\exists_1 v\colon\ v \text{ is in the } x_i\text{-level and } v \text{ represents } g\ ) \Longleftrightarrow g \in \mathrm{dep}(f,q,x_i).$$

EXAMPLE 3.15 *Consider the left BDD given in Figure 3.18. The cofactors in the variables in $q = \{x_1, x_2\}$ are represented by the shaded nodes. Annotated are the sets of variables the cofactors essentially depend on. We have two cofactors depending essentially on $x_3$ and in fact two nodes residing on the third level for the given ordering $\pi^{-1}(x_1) < \pi^{-1}(x_2) < \pi^{-1}(x_3) < \pi^{-1}(x_4)$. With the annotated sets of the left BDD it can be seen already that also two nodes labeled $x_4$ would be situated at the third level for the ordering $\pi^{-1}(x_1) < \pi^{-1}(x_2) < \pi^{-1}(x_4) < \pi^{-1}(x_3)$ since also two cofactors essentially depend on $x_4$. The right BDD then represents the same function using this new ordering, and in fact we have the forecasted number of nodes labeled $x_4$.*

As explained earlier in in Section 3.2.1, $A^*$ requires the heuristic function $h$ as well as the transition cost function $c$ to be well-defined.

The property of well-definedness clearly holds for $h$ by definition. The well-definedness of $c$ is a simple consequence of Lemma 3.14: By construction the transition cost for a transition $q \xrightarrow{x_i} q \cup \{x_i\}$ is $\mathrm{label}(\mathrm{BDD}(f,\pi), x_i) = |\mathrm{dep}(f,q,x_i)|$ where $\pi \in \Pi(q)$ and $\pi(|q|+1) = x_i$. Hence the transition cost function

$$c_{f,n}\colon \{\,t \mid t \text{ is a transition of the form } q \xrightarrow{x_i} q \cup \{x_i\}, \text{ where } x_i \in X_n\,\} \to \mathbb{N}$$

$$c_{f,n}(q \xrightarrow{x_i} q \cup \{x_i\}) = |\mathrm{dep}(f,q,x_i)|$$

is given by a well-defined definition. In the following, the cost of a transition $q \xrightarrow{x_i} q \cup \{x_i\}$ will be denoted $c(q, q \cup \{x_i\})$ as is common in the $A^*$-related literature. We are also omitting the subscripts $f$ and $n$.

[9] Again the quantifier symbol "$\exists_1$" has the semantic "there exists exactly one ...".

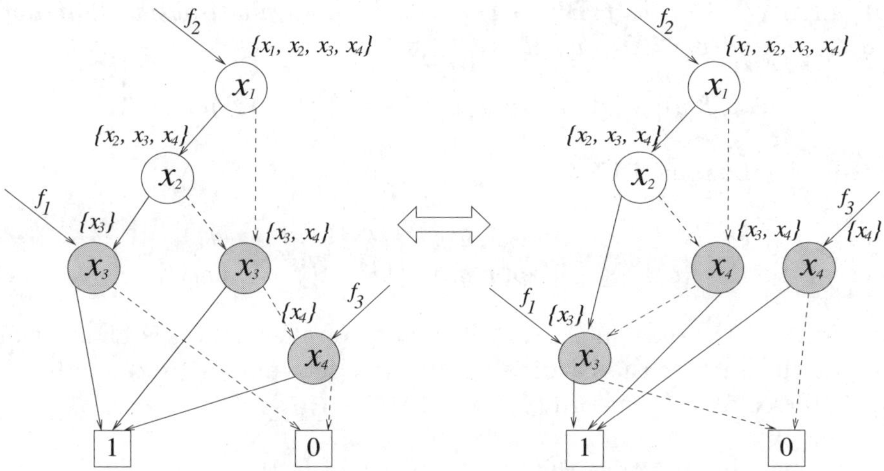


*Figure 3.18.* BDDs for Example 3.15.

The proof of monotonicity is prepared by the next result:

DEFINITION 3.16 *Let $f\colon \mathbf{B}^n \rightarrow \mathbf{B}^m$ be a Boolean multi-output function essentially depending on all variables in $X_n$ and let $x_i \in X_n$. The set of those cofactors of $f$ with respect to variables in $q$ which do not essentially depend on $x_i$ is denoted with* ind$(f, q, x_i)$. *Formally, the function* ind *is given as* ind: $\{f \mid f\colon \mathbf{B}^n \rightarrow \mathbf{B}^m\} \times 2^{X_n} \times X_n \rightarrow 2^{\{f \mid f\colon \mathbf{B}^n \rightarrow \mathbf{B}^m\}}$;

$$\operatorname{ind}(f, q, x_i) = \operatorname{cof}(f, q) \setminus \operatorname{dep}(f, q, x_i).$$

LEMMA 3.17 *Let $f\colon \mathbf{B}^n \rightarrow \mathbf{B}^m$ be a Boolean multi-output function, let $q \subset X_n$ and $x_i \in X_n \setminus q$. It is*

$$\operatorname{cof}(f, q) \subseteq \operatorname{dep}(f, q, x_i) \cup \operatorname{cof}(f, q \cup \{x_i\}).$$

**Proof.** We have

$$\begin{aligned} \operatorname{cof}(f, q) &= \operatorname{dep}(f, q, x_i) \cup \operatorname{ind}(f, q, x_i) \\ &\subseteq \operatorname{dep}(f, q, x_i) \cup \operatorname{cof}(f, q \cup \{x_i\}). \end{aligned} \tag{3.9}$$

It is

$$\operatorname{ind}(f, q, x_i) \subseteq \operatorname{cof}(f, q \cup \{x_i\}) \tag{3.10}$$

since functions in $\operatorname{ind}(f, q, x_i)$ do not essentially depend on $x_i$. Thus, Equation (3.9) holds. □

The following corollary is a direct consequence of Lemma 3.17.

COROLLARY 3.18 *Let $f\colon \mathbf{B}^n \to \mathbf{B}^m$ be a Boolean multi-output function, let $q \subset X_n$ and $x_i \in X_n \setminus q$. It is*

$$|\mathrm{cof}(f,q)| \leq |\mathrm{dep}(f,q,x_i)| + |\mathrm{cof}(f,q \cup \{x_i\})|\,.$$

**Proof.** By Lemma 3.17 it is

$$\begin{aligned} |\mathrm{cof}(f,q)| &\leq |\mathrm{dep}(f,q,x_i) \cup \mathrm{cof}(f,q \cup \{x_i\})| \\ &\leq |\mathrm{dep}(f,q,x_i)| \cup |\mathrm{cof}(f,q \cup \{x_i\})|\,. \end{aligned} \quad (3.11)$$

Equation (3.11) holds since the cardinality of a set union is at most as large as the sum of the cardinalities of the sets united (and smaller if their intersection is non-empty). □

We are now able to prove the monotonicity of $h$.

PROPOSITION 3 *Let $f\colon \mathbf{B}^n \to \mathbf{B}^m$ be a Boolean multi-output function which essentially depends on every variable in $X_n$. The heuristic function $h\colon 2^{X_n} \to \mathbb{N}$*

$$h(q) = \max\left(|\mathrm{cof}(f,q)|\,, n - |q|\right)$$

*is* monotone *with respect to the $A^*$-algorithm given in Section 3.2.2.*

**Proof.** First we assert that in fact it is $h(X_n) = 0$ as required by the definition of heuristic functions. Then we have to show the inequality

$$h(q) \leq c(q, q \cup \{x_i\}) + h(q \cup \{x_i\}) \quad (3.12)$$

for every transition $q \xrightarrow{x_i} q \cup \{x_i\}$ of the algorithm. By construction of the transition cost function described in Section 3.2.1, every state $q$ has been associated with a BDD $F$ having a variable ordering $\pi \in \Pi(q)$. Thus, by Lemma 3.14, Inequality (3.12) can be restated as

$$h(q) \leq |\mathrm{dep}(f,q,x_i)| + h(q \cup \{x_i\}). \quad (3.13)$$

The proof is a careful case analysis of the possible outcomes for the maximum function in the definition of $h$.

**Case 1):**

$$\begin{aligned} |\mathrm{cof}(f,q)| &> n - |q|\,, \\ |\mathrm{cof}(f,q \cup \{x_i\})| &> n - |q \cup \{x_i\}| \end{aligned}$$

Inequality (3.13) follows directly from Corollary 3.18.

**Case 2):**

$$\begin{aligned} |\text{cof}(f,q)| &\leq n - |q|, \\ |\text{cof}(f,q \cup \{x_i\})| &\leq n - |q \cup \{x_i\}| \end{aligned}$$

We have to show

$$n - |q| \leq |\text{dep}(f,q,x_i)| + n - |q \cup \{x_i\}|.$$

We have $|q \cup \{x_i\}| - |q| = 1$ which simplifies the claim to $|\text{dep}(f,q,x_i)| \geq 1$. This clearly holds since $f$ essentially depends on $x_i$ by assumption.

**Case 3):**

$$\begin{aligned} |\text{cof}(f,q)| &> n - |q|, \\ |\text{cof}(f,q \cup \{x_i\})| &\leq n - |q \cup \{x_i\}| \end{aligned}$$

We have to show

$$|\text{cof}(f,q)| \leq |\text{dep}(f,q,x_i)| + n - |q \cup \{x_i\}|.$$

With the assumption $|\text{cof}(f,q \cup \{x_i\})| \leq n - |q \cup \{x_i\}|$ and by Corollary 3.18 we obtain

$$\begin{aligned} |\text{dep}(f,q,x_i)| + n - |q \cup \{x_i\}| &\geq |\text{dep}(f,q,x_i)| + |\text{cof}(f,q \cup \{x_i\})| \\ &\geq |\text{cof}(f,q)|, \end{aligned}$$

yielding the required result.

**Case 4):**

$$\begin{aligned} |\text{cof}(f,q)| &\leq n - |q|, \\ |\text{cof}(f,q \cup \{x_i\})| &> n - |q \cup \{x_i\}| \end{aligned}$$

We have to show

$$n - |q| \leq |\text{dep}(f,q,x_i)| + |\text{cof}(f,q \cup \{x_i\})|.$$

We have $|q \cup \{x_i\}| - |q| = 1$. Furthermore, it is $|\text{dep}(f,q,x_i)| \geq 1$ since $f$ essentially depends on $x_i$ by assumption. With that and with the assumption $|\text{cof}(f,q \cup \{x_i\})| > n - |q \cup \{x_i\}|$ we obtain

$$\begin{aligned} |\text{dep}(f,q,x_i)| + |\text{cof}(f,q \cup \{x_i\})| &> |\text{dep}(f,q,x_i)| + n - |q \cup \{x_i\}| \\ &\geq 1 + n - |q \cup \{x_i\}| \\ &= n - |q|, \end{aligned}$$

yielding the required result. □

Note that by Proposition 2 also the admissibility of $h$ follows. Hence, this is also a formal proof for a property of $h$ informally explained in Section 3.2.1: for the considered BDD and state $q$, $h(q)$ in fact is a lower bound on the cost of the remaining path from $q$ to a goal state, i.e. on the number of nodes in levels $|q| + 1, \ldots, n$.

Moreover, the $\varphi$-values along a path from the initial state to a goal state must be monotonically increasing (in fact, this property yields the intuitive term "monotonicity" here). An example for a monotone heuristic has already been given in Figure 3.15. E.g. for the path $a \longrightarrow b \longrightarrow h \longrightarrow k$, the series $\varphi(a), \ldots, \varphi(k)$ is montonically increasing: $2 \leq 4 \leq 6 \leq 7$.

### 3.2.4 Algorithm

The implementation of the algorithm is based on the implementation of the approaches FizZ and JANUS presented in Section 3.1. Hence, the $A^*$-approach uses the techniques that have been applied there. Among them is the use of symmetries and customized hash tables which are adapted dynamically. Moreover, BDD "explosions" are avoided by undoing variable shifts where appropriate. In Section 3.1, among other aspects, an efficient implementation of the lower bound has been described. Moreover, efficient techniques for fast BDD reconstruction have been proposed here. In the $A^*$-based approach, all these techniques are present or have been further developed. A sketch of the algorithm called $A^{stute}$ is given in Figure 3.19. In the following, a brief comparison of this algorithm to the previous B&B method JANUS is given.

The B&B-paradigm applied in FizZ and JANUS explores the state-space "stage-by-stage", i.e looking at all states with a length $k$ one after the other, then moving on to $k + 1$, etc. (see line (4) in Figure 3.10). Thereby a *static* schema is used to enumerate the states.

$A^*$ however progresses the states in an order based on a *dynamic* evaluation function. The order is not based statically on the length of the state like in the B&B-approach. In Figure 3.19 this is done in line (4), using a counterless "infinite" loop which is exited, if one of two sufficient termination conditions are met (line (12)). A first sufficient condition for termination is that the next state to expand is a goal state, i.e. a state with length $n$. The second sufficient condition is called "Early Termination", a call to the respective macro occurs in line (6). This is a technique to further prune the search space given later in the respective paragraph in Section 3.2.4.1. The code for the technique is given as the macro CHECK-EARLY-TERMINATION in Figure 3.21.

The $A^*$-based approach uses a technique for state expansions which allows to determine transition costs without variable shifts. The computation of the transition costs instead is based on the dependency of cofactors on certain variables, line (17) shows the respective sub-routine call. The whole technique will be explained in detail in Section 3.2.4.5.

Lines (18) to (28) in Figure 3.19 correspond to lines (8) to (20) in Figure 3.10 where the heuristic function $h$ corresponds to the lower bound *lower_bound*, the currently known path cost $g$ corresponds to *minguess*, and the path $p$ corresponds to an ordering $\pi$. However, there are also some differences:

1) No variable shift is needed in algorithm $A^{stute}$ to compute the transitions cost in line (24) in Figure 3.19.

2) No check for already excluded states is done in the $A^*$-approach.

3) A different schema for the update of state data is used, see macro UPDATE-STATE-DATA-ASTAR-TD given in Figure 3.22.

Regarding point 1), this is due to a different technique of state expansion, see Section 3.2.4.5.

Regarding the second point 2): implicitly only states will be expanded that have a potential to lead to an optimal solution, see Theorem 3.19. Therefore, when targeting to reduce the number of state expansions, it would be useless to explicitly exclude states in the $A^*$-based approach. However, it would make sense to use an exclusion mechanism to reduce the number of shifts needed to construct the successor states in an expansion. Such a mechanism in fact will be used in an $A^*$-based bottom-up approach, see Section 3.2.5. In the present top-down approach however, no shifts are performed to construct the successors (see *-1)*), therefore no such exclusion takes place here.

By Proposition 1, the optimal cost for the path to a state is already known at the time of state expansion. This replaces the gradual schema to update the minimum cost used in FizZ or JANUS (see lines (3) to (4) in Figures 3.5 and 3.12). Hence, there also is no need for a lower bound like $l\_b_{cost}$.

Finally, regarding the third point 3), the main differences between UPDATE-STATE-DATA-JANUS (see Figure 3.12) and UPDATE-STATE-DATA-ASTAR-TD (see Figure 3.22) are the following:

a) Only one position of the current ordering must be stored in the $A^*$-based approach (compare line (5) in Figure 3.12 to line (5) in Figure 3.22): this is due to an implementation technique in $A^*$ described in Paragraph "Path Reconstruction" in Section 3.2.4.4. This is an aspect where $A^*$ even needs *less* memory than the B&B-approach.

b) In the $A^*$-based approach, a unique technique to avoid unnecessary computations of the heuristic function and state insertions is applied (see lines (9) and (14)). This technique is called "Delayed State Insertion" and is described in the respective paragraph in Section 3.2.4.1.

The latter point b) as well as the possibility for an early termination is a consequence of a *combination* of $A^*$ and B&B. For this purpose, in line (16) of Figure 3.19 the smallest BDD size seen so far is computed every time to obtain an upper bound. In line (4) in Figure 3.20, the first upper bound is computed as the size of the initial BDD.

In this comparison of the algorithms JANUS and $A^{stute}$ the main differences of the two methods have been outlined. Thereby several new techniques used by the $A^*$-based approach to exact BDD minimization have been mentioned. In the remainder of this section these techniques are described in further detail.

First, a combination of $A^*$ with B&B is described in Section 3.2.4.1. This technique results in a further pruning of the search space as well as a reduction of memory requirement. Moreover, unnecessary computations are avoided. Second, a method for maintaining the set of open states OPEN is given which is faster than the usual heap approach. Next a method to avoid the storing of complete paths is explained which significantly reduces the memory requirement of $A^*$. Note that improvements/implementational aspects of $A^*$ itself are described which can be transferred to other applications using $A^*$-based search techniques. Finally, a unique state expansion technique is suggested that allows to expand states to its successors without the use of variable shifts. This technique yields a significant gain in run time compared to previous approaches to exact BDD minimization in which many time-consuming variable shifts are needed.

#### 3.2.4.1 Combination with Branch and Bound

In this section two techniques to combine $A^*$ with B&B are presented. First, a result from search theory is cited [HNR68, Pea84].

THEOREM 3.19 *Consider a state $q$ in an $A^*$-algorithm operating with evaluation function $\varphi$. ($C^*$ again denotes the minimal cost.) Then we have:*

$$\textit{If } \varphi(q) > C^* \textit{ then state } q \textit{ will not be expanded.}$$

Note that $\varphi(q)$ can change during the algorithm run. In the context of exact BDD minimization, $C^* + 1$ is the minimal BDD size, as $f$, $g$, and $h$ count inner nodes only. Thus, one must be added for the constant node.

```
compute_optimal_ordering(BDD F, int n)
   proc
      INIT-ASTAR
      while 1 do   /* until doomsday */
         determine q ∈ OPEN with minimal
         g[hash(q)] + h[hash(q)];
         CHECK-EARLY-TERMINATION
         if q = X_n then
            reconstruct ordering as the sequence of the path
            to q;
            doomsday:=1;
         end–if
         if doomsday then
            exit while loop;
         end–if
         reconstruct an appropriate ordering π ∈ Π(q) with
         path reconstruction;
         establish π on F;
         upper_bound := update_upper_bound();
         compute dependencies(F, |q|);
         for each x_i ∈ X_n \ q do
            CHECK-SYMMETRY
            q' := q ∪ {x_i};
            if q' ∉ states then
               g[hash(q')] := ∞;
            end–if
            newcost := label(F, |q'|) + g[hash(q)];
            UPDATE-STATE-DATA-ASTAR-TD
            UNDO-SHIFT
            end–if
         end–for
      end–while
   end–proc
```

*Figure 3.19.* Algorithm $\boldsymbol{A}^{stute}$.

In a B&B algorithm to determine a minimum (e.g. minimal cost), a lower and an upper bound on the minimum are constantly updated during the algorithm run. This happens every time new information is available to tighten the bounds. Next, the $A^*$-approach to exact BDD minimization is *combined* with B&B by continously updating an up-

```
INIT-ASTAR
    macro
        π[hash(∅)] := an arbitrary initial order;
        upper_bound := update_upper_bound();
        g[hash(∅)] := 0;
        lower_bound[hash(∅)] := number of output nodes m;
        states[hash(∅)] := ∅;
        insert ∅ into OPEN;
        doomsday := 0;
    end–macro
```

*Figure 3.20.* Initialization for the $A^*$-based approach.

```
CHECK-EARLY-TERMINATION
    macro
        if g[hash(q)] + h[hash(q)] + 1 = upper_bound then
            reconstruct ordering that yielded upper_bound;
            doomsday:=1;
        end–if
    end–macro
```

*Figure 3.21.* Early termination.

per bound (denoted *upper_bound*) during the algorithm run every time a smaller BDD size is found.

**Delayed State Insertion.** The integration of B&B into $A^*$ is based on the following observation: Clearly, we have

$$upper\_bound \geq C^* + 1. \tag{3.14}$$

Now let $q$ be a state with $\varphi(q) + 1 > upper\_bound$. With (3.14) we have

$$\varphi(q) + 1 > C^* + 1$$

and thus, by Theorem 3.19, $q$ will not be expanded by $A^*$.

In the $A^*$-based approach, this result is used as follows: a state $q$ is inserted into OPEN iff $\varphi(q) + 1 \leq upper\_bound$ (see line (14) in Figure 3.22). Otherwise, the insertion is delayed until $g(q)$ has become small enough (by continuous updates during the algorithm run) to meet this condition. This strategy avoids the inclusion of many states into OPEN

```
(1)  UPDATE-STATE-DATA-ASTAR-TD
(2)    macro
(3)      if q' ∉ states or newcost < g[hash(q')] then
(4)        g[hash(q')] := newcost;
(5)        p(q') := x_i;
(6)        if q' ∉ states then
(7)          states[hash(q')] := q';
(8)        end–if
(9)        if h[hash(q')] not yet computed and
           g[hash(q')] + n − |q'| + 1 ≤ upper_bound then
(10)         shift x_i to level |q| + 1;
(11)         upper_bound := update_upper_bound();
(12)         h[hash(q')] := compute_lower_bound(q');
(13)       end–if
(14)       if h[hash(q')] computed and
           g[hash(q')] + h[hash(q')] + 1 ≤ upper_bound then
(15)         if OPEN too crowded, resize it and
             garbage-collect outdated entries;
(16)         insert q' into OPEN;
(17)       end–if
(18)     end–if
(19)   end–macro
```

*Figure 3.22.* Updating the state data in the $A^*$-based top-down approach.

which never will be expanded. This reduces the memory requirement of the algorithm. Note that in line (9) in Figure 3.22 also the computation of $h(q)$ is delayed to save unnecessary computations: by definition $n - |q| \leq h(q)$, thus we have $g(q) + n - |q| \leq \varphi(q)$. Consequently, $g(q) + n - |q| + 1$ can be tested instead of $\varphi(q) + 1$ for being less or equal *upper_bound*. If this simple test already fails, the failure of the "full" test $\varphi(q) + 1 \leq \mathit{upper\_bound}$ is implied, hence the computation of $h(q)$ can be delayed. That way, it is possible that for the state $q$, $h$ is never needed to compute, avoiding unnecessary computations.

**Early Termination.** Integrating B&B also allows to further prune the state space, based on the following consideration: Let $q$ be a state which is chosen as the next best state, i.e. $q$ will be expanded. Inversion (and adding one to each side of the resulting inequality) of the statement in Theorem 3.19, together with Equation (3.14) yields

$$\varphi(q) + 1 \leq C^* + 1 \leq \mathit{upper\_bound}. \tag{3.15}$$

Now let $\varphi(q) + 1 = upper\_bound$. We have $upper\_bound = C^* + 1$ by Inequation (3.15), i.e. the upper bound has already reached the minimum.

In the $A^*$-based approach, this result is used as follows: if the total cost of the state currently considered (plus one, for the constant node) already equals the upper bound, the algorithm is finished by reconstructing the variable ordering which yielded this upper bound (see lines (3)-(6) in Figure 3.21). This "early termination" results in a further pruning of the search space.

#### 3.2.4.2 Maintaining the Set of Open States

In this section an efficient implementation of the priority queue OPEN is described which uses a customized data structure instead of the standard heap implementation used by previous approaches. As described in Section 3.2.1, OPEN, the set of "open" states, implements the set of states that the algorithm still considers as candidates for state expansion. The only operations applied to OPEN are deletions of elements with smallest total cost, i.e. most promising states according to Equation (3.3) and insertion of elements, i.e. newly generated successor states. Thus OPEN is essentially a priority queue which is often implemented as a heap, e.g. see [Nil80] or [KRR88]. The heap implementation requires insertions and deletions in $O(\log N_{\text{OPEN}})$ steps where $N_{\text{OPEN}}$ is the size of OPEN [AHU74]. As explained in the previous sections, some (or even most) of the states contained in OPEN may eventually be expanded, i.e. we have $N_{exp} \leq N_{\text{OPEN}} \leq N_{space}$ where $N_{exp}$ denotes the number of state expansions and $N_{space}$ denotes the size of the state space.

To obtain a worst-case analysis, the introduced quantities are all assumed to be of the same order:

$$O(N_{exp}) = O\left(N_{\text{OPEN}}\right) = O(N_{space}) \tag{3.16}$$

In the application investigated here, exact BDD minimization, the size of the state space is

$$N_{space} = 2^n,$$

i.e. it grows exponentially with $n$ where $n$ is the number of input variables of the Boolean function represented by the BDD to minimize. Thus, by Equation (3.16), insertion and deletion operations of the heap implementation still show a run time complexity of $O(n)$. In the $A^*$-based approach discussed, an even more sophisticated technique has been chosen to implement the set OPEN. This technique makes use of a theoretical result stating that the sequence of minimal costs for the states chosen to expand is monotonically increasing [Pea84, HNR68]. This property allows a fast strategy to lookup the next minimum with an only quasi-constant effort per expansion. Note that the idea of this technique can be transferred to other applications using the $A^*$-algorithm as well: for some applications, $N_{space}$ even grows proportional to $n!$ where $n$ denotes the size of the problem instance, i.e. it grows faster than exponentially. Then, by Equation (3.16) also $N_{\text{OPEN}}$ and $N_{exp}$ grow that fast. Examples are the TSP (Traveling Salesman Problem) and other problems of combinational optimization. Here, the discussed heap operations, encountered for every state expansion, may significantly slow down run time and a quasi-constant method is strongly desirable. Next, after citing a result from search theory [HNR68, Pea84], the new method and the underlying data structure are described.

PROPOSITION 4 *Consider an $A^*$-algorithm operating with evaluation function $\varphi$. Let $q$ be the state which least recently was expanded and let $q'$ be the state chosen by the algorithm to be expanded next. Then we have*

$$\varphi(q') \geq \varphi(q).$$

In addition to this result, we have $\varphi(q) \leq C^*$ for all states $q$ being expanded as a consequence of Theorem 3.19. Thus, the sequence of cost minima is *monotonically increasing* within the range $\varphi(q_0), \dots, C^*$ where $q_0$ is the initial state. For the application here (and for many other AI applications as well[10]) this range is an interval which is very small compared to the number of expansions during the algorithm run $N_{exp}$, i.e. we have

$$C^* \ll N_{exp} \in O(N_{space}) = O(2^n). \tag{3.17}$$

This result can be used in the following way: Let $C'$ be the size of the BDD initially given to the algorithm. An array OPEN of size $C'$ of (initially empty) state-lists is allocated. Clearly we have $C' \geq C^*$, hence the subscripts of the array cover the whole range of possible cost minima. The idea is to append each newly generated state $q$ to the list OPEN$[\varphi(q)]$.

In the beginning, $q_0 = \emptyset$ is appended to the list OPEN$[m]$ (which initially is empty) where $m$ is the number of output nodes. The algorithm starts expanding state $q_0 = \emptyset$ with currently minimal cost $\varphi(\emptyset) = g(\emptyset) + h(\emptyset) = 0 + m = m$. $m$ is stored as the previous minimum

---

[10] E.g., for the TSP, $C^*$ is the length of an optimal tour which typically can be bounded by $c \cdot n$ where $n$ is the number of cities and $c$ is some (small) constant, e.g. the longest distance between two cities. Thus, for the TSP there is a similar relation $C^* \leq c \cdot n \ll N_{space} = n!$

*prev_min* for the next expansion and $q_0$ is deleted from OPEN$[m]$. Moreover, the newly generated successors of $q_0$, say $q_1, \ldots, q_k$, are inserted into OPEN by appending $q_i$ to list OPEN$[\varphi(q_i)]$. The same schema of state insertion is used for every successor state generated throughout the algorithm.

Whenever a new state with minimal cost on OPEN must be determined for expansion, the range $prev\_min, \ldots, C'$ is scanned by repeatedly increasing a counter $cnt$ each step by one from $prev\_min$ to $C'$, until the next non-empty list OPEN$[cnt]$ is found. Then the state least recently appended (i.e. the last state of the list OPEN$[cnt]$) is chosen as the next state $q$ to expand. As in the beginning for $q_0$, after storing $\varphi(q)$ as the next previous minimum, $q$ is deleted from OPEN$[\varphi(q)]$.

With the result from Proposition 4, this strategy must be correct since always a state with the next smaller $\varphi$-value is determined in the range of the monotonically increasing series of minima.

Note that adding and deleting elements (i.e. states) from OPEN requires only constant time. Determining the next state to expand is an operation with a total time effort for *all* expansions during the whole algorithm run which is proportional to $C^*$. This holds since in one run of the $A^*$-algorithm the array OPEN is only scanned once starting from OPEN$[\varphi(q_0)]$ to OPEN$[C^*]$. Testing a list for non-emptiness each scanning step is done in constant time by maintaining appropriate element counters. By these considerations, the total time effort for the maintainment of OPEN is $O(N_{exp} + C^*)$ (the list OPEN is accessed $N_{exp}$ times during the expansions of the algorithm). This contrasts to an effort of $O(N_{\text{OPEN}} \cdot \log N_{\text{OPEN}})$ in the case of a heap-based maintainment of OPEN (note that $O(N_{\text{OPEN}}) = O(N_{exp})$ by Equation (3.16)). With Inequality (3.17) we have $O(N_{exp} + C^*) = O(N_{exp})$, thus the strategy described can yield significant improvements for all application domains where a relation similar to Inequality (3.17) can be observed.

#### 3.2.4.3 Maintaining Closed States

In its general description, $A^*$ selects a most promising state $q$ (i.e. the state with the smallest $\varphi$-value) from OPEN for expansion, generates its successors and inserts the successors into OPEN. The state $q$ is inserted into a list of so-called "closed" states CLOSED. This list is maintained for the following reason: there can be more than one path by which a particular state can be reached from the initial state (see state $i$ in Figure 3.23, depicting the last step of the example given on page 70). Cheaper paths can be found *after* state $q$ has been closed. In the general case, the heuristic function $h$ can be non-monotone. Thus, $q$ may be reopened during further progress of the algorithm since Theorem 3.12

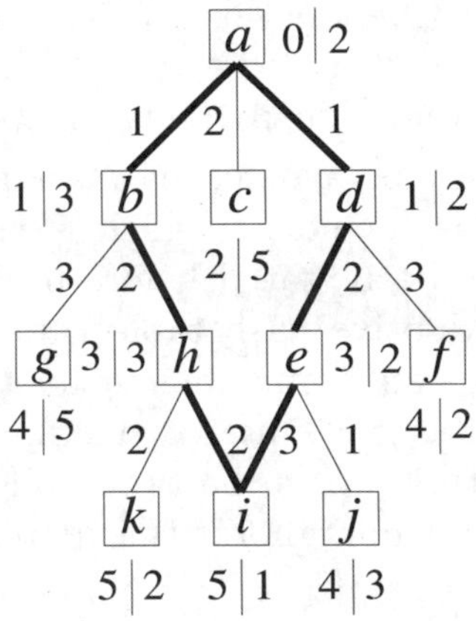


*Figure 3.23.* Two paths leading to state $i$.

does not apply in this case. Consequently, states still need to be stored after expansion in order to be able to update path costs each time a cheaper path is found.

Additionally, the step of inserting the successors into OPEN is more complicated since it also has to be checked for cheaper paths found to states already contained in CLOSED, e.g. see [Pea84].

However, in the case of a *monotone* function $h$, maintaining closed states is much simpler, using the result from Proposition 1 [HNR68, Pea84]: this result states that $A^*$, guided by a monotone heuristic, finds *optimal* paths to all expanded states. In other words, no improved paths to a state $q$ can be found after $q$ has been expanded. Consequently, no code managing path cost updates for the states in CLOSED is necessary in the algorithm. This significantly simplifies the situation.

Can we do *without* a list CLOSED then? With the previous consideration, the answer seems to be "yes". However, the following situation has to be considered: Let $q$ be a state which already has been expanded. Assume a path from the initial state to $q$ *worse* than the best (i.e. cheapest) one known is found after expansion of $q$. This is possible: Proposition 1 does not state that *all* paths to $q$ have been considered at the time of expansion. Proposition 1 only states that the paths considered so far must already include the best one.

If the algorithm simply deleted $q$ from OPEN, state $q$ would be treated as a newly encountered state. Consequently the method inserts $q$ into OPEN again. This is because there is no way to detect that $q$ already has been expanded before. Even worse, since the algorithm now works with a $\varphi$-value for $q$ *higher* than the (minimal) $\varphi$-value $q$ showed at the time of expansion, $q$ may in fact be identified again as minimal on OPEN and thus $q$ might be mistakenly expanded again. This is a consequence of the series of minima being monotonically increasing (see Section 3.2.4.2).

This situation can easily be avoided by marking $q$ as "closed" in the usual adaptive hash table for states which is used by the framework of the algorithm. As has been stated before in Section 3.1, basing the implementation on an adaptive hash table (first suggested in [JKS93, DDG00]) means an important improvement for the general framework. Whenever a closed state is encountered again, the theory guarantees that the newly found path is worse than the best one found at the time of expansion and the algorithm ignores this state.

#### 3.2.4.4 Reconstruction Techniques

This section describes an implementation technique of the $A^*$-based approach to exact BDD minimization which significantly reduces the space requirement of the $A^*$-algorithm. This is an effective remedy for the large space demand of $A^*$ explained in Section 3.2.1.

In the $A^*$-based algorithm for exact BDD minimization, a path from the initial state via a state $q$ is described as a sequence of input variables. By construction the BDD representing state $q$ respects a variable ordering $\pi \in \Pi(q)$ (see Equation (3.7) in Section 3.2.2). A simple version of $A^*$ would store the currently cheapest known path to $q$ in a state attribute $p(q)$. With the optimization described in this section, this is not necessary.

All previous algorithms for exact BDD minimization store complete variable orderings, as BDDs have to be *reconstructed* frequently during an algorithm run. This is because storing (and reusing) complete BDDs instead would cause an intolerable memory overhead.

As mentioned above, the modified version of the $A^*$-algorithm does not store complete orderings. Therefore, a special BDD reconstruction technique is necessary which is given in Paragraph "BDD Reconstruction" of this section. This technique turns out to be significantly faster than previous BDD reconstruction techniques in exact BDD minimization.

**Path Reconstruction.** In algorithm $A^{state}$, a large reduction in memory requirement is achieved by the following strategy: Instead of storing and continously updating the whole sequence of the cheapest known path from the initial state to a state $q$, only one variable is stored in $p(q)$. This is the variable denoted at the transition of the predecessor to $q$ which yielded the cheapest path to $q$ known so far.

In the application of an exact BDD minimization algorithm, this saves a large amount of memory since the memory requirement for $p(q)$ now is only $O(\log n)$ for one variable identifier instead of $O(n \cdot \log n)$ where $n$ is the number of input variables. Moreover, the previous requirement of $O(n \cdot \log n)$ was the largest for all state attributes. A state $q$ itself can be encoded in space $O(n)$ and $g(q)$, $h(q)$ are natural numbers requiring space $O(\log n)$.

Of course, the full sequence of a cheapest known path to a state $q$ must be known at least at termination of an $A^*$-algorithm, as the optimal path to a goal state is the solution computed by $A^*$. In the $A^*$-based exact BDD minimization algorithm, the full sequence will be needed even more frequently during the algorithm run, as is explained in the following section.

The cheapest known path can be reconstructed as follows: the last position in the sequence of variables along this path must be $p(q)$ since this was the variable annotated at the last transition to $q$. This transition was

$$r := q \setminus \{p(q)\} \xrightarrow{p(q)} q.$$

Hence, with an analogous argument, the position before the last position must be $p(r)$ etc. In this, in a kind of "reverse construction", the last up to the first position of the ordering along the cheapest known path can be reconstructed one after the other.

Assuming a good hashing function designed to access $p(q)$ as entry in a hash table via key $q$, i.e. assuming quasi-constant access behavior, the run time complexity of the described path reconstruction is only $O(n)$ where $n$ is the number of input variables of the BDD. Hence, this operation has minor time complexity which is negligible compared to other dominating operations carried out with each expansion (e.g. variable shifts).

This section is finished with a formal proof that the above technique is correct, i.e. soundness of the modified $A^*$-algorithm is preserved.

LEMMA 3.20 *Consider an $A^*$-algorithm with the modification that for each state $q$ only the label of the last transition which was yielding a decrease of $g(q)$ is stored in $p(q)$ during the algorithm run. Then the full sequence $\pi(1), \ldots, \pi(|q|)$ of a cheapest known path from the initial state to a state $q$ can be reconstructed as*

$$\pi(i) = p(q \setminus \bigcup_{j=i+1}^{|q|} \{\pi(j)\})$$

*for $1 \leq i \leq |q|$.*

That is, starting with $\pi(|q|) = p(q)$, the full sequence $\pi(1), \ldots, \pi(|q|)$ can be iteratively reconstructed going from right to left, using all previously constructed rightmost elements in this sequence.

**Proof.** The proof is an induction on $k := |q|$. First, let $k = 0$. Then $q$ must be the initial state. The only path from the initial state to $q$ is the empty sequence which clearly is cost minimal, i.e. the cheapest known path. Now assume the result holds for $k$ and let $|q| = k + 1$. W.l.o.g. let $p(q) = x_m$. Then $r := q \setminus \{x_m\}$ is the predecessor of $q$ (i.e. $r \xrightarrow{x_m} q$) on the cheapest known path from the initial state to $q$. Let $\pi_r$ be the cheapest known path from the initial state to $r$. It is $|r| = k$, hence by induction hypothesis, $\pi_r$ can be reconstructed and it is

$$\pi_r(i) = p(r \setminus \bigcup_{j=i+1}^{|r|} \{\pi_r(j)\})$$

and since $r$ is the predecessor of $q$ on the cheapest known path for $q$ and, as the key observation, *since a cost-minimal path must consist of cost-minimal sub-paths*, we have

$$\pi_q(i) = p((q \setminus x_m) \setminus \bigcup_{j=i+1}^{|q|-1} \{\pi_q(j)\}) \tag{3.18}$$

for $1 \leq i \leq |q| - 1$. Now the cheapest known path to $q$ can be reconstructed by simply appending the variable $x_m$ to the sequence of variables in $\pi_r$. Consequently, we also have $\pi_q(|q|) = x_m = p(q)$, yielding the claimed result together with Equation (3.18). □

**BDD Reconstruction.** As all exact BDD minimization algorithms suggested so far, the $A^*$-based approach needs to reconstruct BDDs respecting a certain variable ordering: prior to expansion of a state $q$, a BDD respecting an ordering $\pi \in \Pi(q)$ is reconstructed in line (14) of algorithm $A^{stute}$ (see also Equation (3.7) in Section 3.2.2). The first $|q|$ positions of such an ordering are obtained with the path reconstruction technique of the previous section. The remaining last $n - |q|$ positions can in principle be chosen arbitrarily.

Unfortunately, this sometimes may result in an unacceptable large increase of the size of the lower part of the BDD. In Section 3.1.3.2, discussing the BDD reconstruction method applied in JANUS, a remedy for this effect was suggested which tries to avoid moving away from good orderings: whenever the size of the reconstructed BDD exceeds the original size of the BDD by a certain factor, the remaining positions

$\pi(|q+1|), \ldots, \pi(n)$ are chosen from the old ordering of the reconstructed BDD. This was possible as previous approaches stored complete variable orderings.

As complete orderings are not stored in the $A^*$-based approach (as explained in the previous section), a more sophisticated technique for this must be found. The $A^*$-based method is directly oriented towards an ordering that is already known to be good. The experiments show that this is more effective, see Section 3.2.6.

In the following the former approach of Section 3.1.3.2 is compared to the new BDD reconstruction technique given here in detail. For this, first the reason behind the effectiveness of the technique used in JANUS is identified. By this more insight into the problem is gained. Second, a different method is suggested based on the new view of the problem.

The following holds for all exact BDD minimization algorithms published so far: at the start of the algorithm, a good initial ordering is determined by Rudell's algorithm for dynamic reordering (see Section 2.4.7.2). This ordering changes during the algorithm run by the shifting of variables. These shifts either extend orderings by one (next) variable or states are expanded (as in the $A^*$-based approach to exact BDD minimization). In top-down approaches like the method presented here, variables are shifted one after the other from the lower part to the upper part of the BDD. In this, (with an additional assumption) the following holds:

*The relative order of the variables in the lower part remains the same as in the initial ordering.* (3.19)

EXAMPLE 3.21 *Let the initial ordering $\pi$ be*

$$\pi(1) = x_1, \pi(2) = x_2, \pi(3) = x_3, \pi(4) = x_4, \pi(5) = x_5.$$

*Let $F$ be a BDD respecting $\pi$ and let the upper part consist of the first two BDD-levels. Therefore, the lower part consists of the last three BDD levels. Then, the relative ordering in the lower part is $\pi^{-1}(x_3) < \pi^{-1}(x_4) < \pi^{-1}(x_5)$.*

*If $x_3$ is shifted to the second level, i.e. the bottommost level of the upper part, the resulting ordering is*

$$\pi(1) = x_1, \pi(2) = x_3, \pi(3) = x_2, \pi(4) = x_4, \pi(5) = x_5.$$

*The number of variables in the lower part has decreased by one and the relative ordering now is $\pi^{-1}(x_4) < \pi^{-1}(x_5)$ for the variables $x_4$ and $x_5$ in the lower part as well as $\pi^{-1}(x_4) < \pi^{-1}(x_5)$ in the initial ordering.*

In detail, the situation is as follows: as is common for all recent methods, all variables in the lower part are shifted to the bottommost position in the upper part. These shifts provide all possible extensions of the ordering (in the B&B-based approaches) or expansions of the current state (in the $A^*$-based approach). After that, the upper part has increased by one variable and the above is repeated. The above Invariance (3.19) only holds if the following assumption is met: the shifts to the same (bottommost) position must be *undone*, i.e. a variable which was shifted up must be shifted back to its old position before choosing the next variable to shift (see also the algorithms FizZ and JANUS in Section 3.1). That way, the original situation is reestablished before each next shift. Otherwise the relative order of the initial ordering is destroyed more and more for the variables in the lower part by every variable shift.

EXAMPLE 3.22 *If $x_4$ is shifted to the second level after having shifted $x_3$ in the above example, the resulting ordering is*

$$\pi(1) = x_1, \pi(2) = x_4, \pi(3) = x_3, \pi(4) = x_2, \pi(5) = x_5.$$

*If now $x_5$ is shifted to the second level, the ordering*

$$\pi(1) = x_1, \pi(2) = x_5, \pi(3) = x_4, \pi(4) = x_3, \pi(5) = x_2$$

*is obtained. In the lower part, the relative ordering is $\pi^{-1}(x_3) < \pi^{-1}(x_2)$ whereas it is $\pi^{-1}(x_2) < \pi^{-1}(x_3)$ in the initial ordering.*

That way, the algorithm stays as close as possible to an example of a good ordering: the initial ordering. The risk of BDD explosion in the lower part is kept minimal and non-accumulative, i.e. small increases in size are not propagated to other BDDs and hence do not sum up to a larger increase.

In the upper part a new order of the variables is established gradually. Bad orderings in the upper part yield large sizes and hence soon stop that ordering from being further examined, either due to exclusion by bounds (as in B&B methods) or by the evaluation of a cost function in $A^*$. Hence, explosions in the upper part can be expected not to be propagated very far to other BDDs during the algorithm.

Summarized: if BDDs are reconstructed to their old ordering, i.e. the ordering they had achieved from a previous step (as outlined above), Invariance (3.19) holds. By this, moving away from good orderings is avoided.

Unfortunately, the previous assumption that every shift to the bottommost position is undone cannot be met in practice since this involves too many additional, time-consuming variable shifts (see also the description of the undo-operation in the macro given in Figure 3.6 in Section 3.1). Hence, in all recent approaches shifts are undone only, if the size of the BDD increased too much by shifting. Hence, the strategy described in Section 3.1.3.2 will fail to address the true cause of the BDD explosion problem if too many shifts are not followed by an undo.

Now that more insight into the nature of the problem has been gained, a better solution is almost obvious: at state $q$, instead of returning to the old ordering of the BDD to reconstruct for $q$, the ordering of the variables in the lower part is *directly* set such that the same relative ordering of these variables as in the *initial ordering* is established. Therefore, the initial ordering is stored at start of the method to keep it available during the whole algorithm run. Since the old ordering of the reconstructed BDD does not need to be known, this method can easily be applied together with the path reconstruction method described in the previous section.

#### 3.2.4.5 State Expansion Technique

In this section a unique state expansion technique is described which determines the cost of the resulting transitions without involving time consuming variable shifts.

Suppose a BDD for a Boolean function $f$ is minimized. When a state $q$ is expanded in algorithm $A^{state}$, for all successors $q' = q \cup \{x_i\}$ ($x_i \in X_n \setminus q$) the costs for the transitions $q \xrightarrow{x_i} q \cup \{x_i\}$ must be computed. By the construction described in Section 3.2.2 this reduces to computing the terms label(BDD$(f, \pi_i), x_i)$ where $\pi_i$ is a variable ordering such that $\pi_i \in \Pi(q)$ and $\pi_i(|q'|) = x_i$. A trivial approach to this is the actual construction of the BDDs BDD$(f, \pi_i)$ requiring the shifting of the variable $x_i$ to the $(|q|+1)$-th level. This is a very time consuming operation with the potential risk of increasing the number of BDD nodes ("BDD explosion"). The approach given here applies a more sophisticated technique for this, using an argument which follows Lemma 3.14 in Section 3.2.3. Reconsider Example 3.15: With the annotated sets of the left BDD in Figure 3.18 it can be seen *without actually shifting a variable* that also two nodes labeled $x_4$ would be situated at the third level for the ordering

$$\pi^{-1}(x_1) < \pi^{-1}(x_2) < \pi^{-1}(x_4) < \pi^{-1}(x_3).$$

This argument is used in algorithm $A^{state}$ in Figure 3.19, line (17) as follows: a BDD $F$ for $q$ with an ordering $\pi \in \Pi(q)$ can be assumed by construction (see above and Section 3.2.2). Computing the terms label(BDD$(f, \pi_i), x_i)$ for all $x_i \in X_n \setminus q$ simplifies to counting those cofactors with respect to a variable in $q$ which essentially depend on $x_i$. To prepare this, a bottom-up traversal in the levels $|q| + 1, \ldots, n$ associates with each node a set of those variables which the function represented by the node essentially depends on. The sets annotated at the left BDD in Figure 3.18 demonstrate the idea of this construction: at a node which is testing a variable $x_i$, the following interrelation derived from the Shannon decomposition is used:

$$\text{support}(f) = \{x_i\} \cup \text{support}(f_{x_i=1}) \cup \text{support}(f_{x_i=0})$$

where support$(f)$ again denotes the set of variables in the support of $f$, i.e. the set of variables $f$ essentially depends on. The sets are implemented as bit-sets, each bit corresponding to the subscript (index) of a variable (see variable $q$ in line (5) in Figure 3.24, $q$ is a mask of $n$ bits). In line (6), the $i$-th bit is set in mask $q$ if var$(v) = x_i$. Hence, the union of these sets can be implemented with a simple bitwise or-operation (see line (8)). After construction of these sets, in top-down manner, the required cofactors are identified and counted *all at once* in *one* traversal (increasing counters for every variable if a cofactor essentially depends on that variable, see array *label* in Figure 3.24). All this can be done with a procedure traversing the nodes (twice) in the levels $|q|+1, \ldots, n$ which does not involve any variable shifts (see Figure 3.24).

### 3.2.5 Bottom-Up Approach

In the previous sections, an $A^*$-based approach to exact BDD minimization has been described which should be called a *top-down* approach for the following reason: a state $q$ always is associated with a BDD whose *first* $|q|$ variables are ordered according to the *first* $|q|$ positions of the variable ordering which is annotated at the transitions of the currently considered path to $q$ (see Section 3.2.2). There is also vital interest in *bottom-up* approaches: in Section 3.1.4, experiments have been reported which show that for arithmetic functions like multipliers a bottom-up approach can be faster than a top-down approach by one order of magnitude. Moreover, also some benchmark test-cases indicated exceptions from the general rule that top-down approaches are superior to their bottom-up counterparts. Similar results were reported in [DDG00].

The conclusion is that e.g. running both a good bottom-up and top-down approach in parallel on different machines can help to find a solution faster. This is an industrial standard practice often applied if for a given problem the best of several available algorithms is not known in advance. Hence, the added flexibility of an approach which can be applied both top-down and bottom-up would be greatly desirable.

In this section an extension of the $A^*$-based approach to exact BDD minimization is described which can be applied bottom-up. The basic framework remains the same with the following differences: instead of considering the first $|q|$ variables and positions of the variable orderings

```
compute_dependencies(BDD F, int level, int n)
    proc
        for i := n downto level + 1 do
            for each node v in level i do
                q := 0;
                set the (varindex(v))-th bit in q;
                if i < n then
                    set all bits in q which are set in
                    masks[hash(then(v))] or masks[hash(else(v))];
                end–if
                masks[hash(v)] := q;
            end–for
        end–for
        set all elements of the array label to 0;
        for i := level + 1 to n do
            for each node v in level i representing a cofactor
            with respect to the first level variables do
                for j := 0 to n − 1 do
                    q := masks[hash(v)];
                    if the j-th bit is set in q then
                        label[j] := label[j] + 1;
                    end–if
                end–for
            end–for
        end–for
    end–proc
```

*Figure 3.24.* Computing the dependencies.

annotated along the currently considered path to $q$, always the *last* $|q|$ variables and respective positions are chosen.

This slightly changes the transition cost function: the cost of a transition $q \xrightarrow{x_i} q \cup \{x_i\}$ now is

$$c_{f,n}(q \xrightarrow{x_i} q \cup \{x_i\}) = |\mathrm{dep}(f, X_n \setminus (q \cup \{x_i\}), x_i)|\,.$$

However, the required well-definedness is preserved. Another difference lies in the use of the heuristic function $h$: let $(f_i)_{(1 \le i \le m)}$ be the family of single-output functions constituting a Boolean multi-output function $f$: $\mathbf{B}^n \to \mathbf{B}^m$. Let r_upper: $X_n \to \mathbb{N}$; r_lower: $X_n \to \mathbb{N}$; be defined as

$$\text{r\_upper}(q) \quad = \quad |\{f_i \mid \exists x \in q\colon\ f_i \text{ essentially depends on } x\}|$$

and

$$\text{r\_lower}(q) = |\{ f_i \mid \nexists x \in q\colon f_i \text{ essentially depends on } x \text{ and } f_i \notin \text{cof}(f, q) \}|.$$

Now let roots: $X_n \to \mathbb{N}$; $\text{roots}(q) = \text{r\_upper}(q) + \text{r\_lower}(q)$. The bottom-up approach uses the heuristic function

$$h(q) = \max\left(|\text{cof}(f, X_n \setminus q)| - \text{roots}(X_n \setminus q), n - |q|\right).$$

First, some motivation is given to this notation and definitions: Let $q \subseteq X_n$ and consider a BDD $F$ representing $f$ and respecting a variable ordering $\pi$ with $\pi \in \Pi(q)$. The term r_upper($q$) ("upper outputs") denotes the number of single output functions which are represented by a node in the upper part of $F$, i.e. represented by a node labeled with a variable in $q$. The term r_lower($q$) ("lower outputs") denotes the number of single output functions which are represented by a node in the lower part of $F$, i.e. they are represented by a node labeled with a variable in $X_n \setminus q$, and whose representing nodes do not already represent a cofactor with respect to the variables in $q$. The idea behind the last definition is not to count functions or the respective nodes twice, as the lower bound used, $h(q)$, already counts the functions (or the representing nodes, respectively) in $\text{cof}(f, X_n \setminus q)$.

The function $h$ can be interpreted as an acceleration of the well-known lower bound given in the exact minimization algorithm of [ISY91]: this bottom-up B&B algorithm uses the number of nodes situated at the $(n - |q| + 1)$-th level decreased by the number of output nodes[11]. The idea is to calculate a minimal number of nodes in the part of the BDD above level $n - |q| + 1$ to connect the root nodes to the nodes of level $n - |q| + 1$.

By Lemma 3.14, the nodes in level $n - |q| + 1$ represent cofactors with respect to the first $n - |q|$ variables of the ordering, i.e. they are a *subset* of $\text{cof}(f, X_n \setminus q)$. With that the heuristic function $h$ proposed here for a bottom-up approach is *tighter* than that used in [ISY91].

How can it be seen that $h(q)$ is a lower bound on the minimal number of nodes in levels $1, \ldots, |q|$ then? For now it can simply be remarked that an idea similar to Lemma 3.5 and Corollary 3.6 applies. No new proof for this is given here since in the next section a stronger result will be proven, namely the monotonicity of $h$. With the result of Proposition 2 in Section 3.2.3, the required result, the admissibility of $h$, directly follows.

[11] In fact, in [ISY91] the bound is actually given only for the case of a BDD representing a single output function, hence one output node is substracted.

#### 3.2.5.1 Monotonicity

The proof of the monotonicity of $h$ is prepared with the following two results. Let roots, r_upper and r_lower be defined as above.

LEMMA 3.23 *For $q \subset q' \subseteq X_n$ we have* $\text{roots}(q) \leq \text{roots}(q')$.

**Proof.** It is $\text{r\_upper}(q) + \text{r\_lower}(q) \leq \text{r\_upper}(q') + \text{r\_lower}(q')$ since "lower outputs" on the left side of the inequality which are *not* always counted by definition, turn to "upper outputs" on the right side which are always counted. □

LEMMA 3.24 *Let $f: \mathbf{B}^n \to \mathbf{B}^m$ be a Boolean multi-output function, let $q \subseteq X_n$ and $x_i \in X_n \setminus q$. It is*

$$|\text{cof}(f, q \cup \{x_i\})| \leq |\text{dep}(f, q, x_i)| + |\text{cof}(f, q)|\,.$$

**Proof.** Functions contained in $\text{cof}(f, q \cup \{x_i\})$, but not contained in $\text{ind}(f, q, x_i)$, must be direct cofactors in $x_i$ of functions in $\text{cof}(f, q)$ that are essentially depending on $x_i$. Formally, we have

$$\begin{aligned} dcof &:= \left\{ g|_{x_i=b} \mid g \in \text{dep}(f, q, x_i) \text{ and } b \in \mathbf{B} \right\} \\ &\supseteq \text{cof}(f, q \cup \{x_i\}) \setminus \text{ind}(f, q, x_i). \end{aligned}$$

We have $\text{ind}(f, q, x_i) \subseteq \text{cof}(f, q \cup \{x_i\})$ by Equation (3.10) in Section 3.2.3. Hence, the cofactor set $\text{cof}(f, q \cup \{x_i\})$ can be partitioned into two disjoint sets, $\text{ind}(f, q, x_i)$ and some set $subdcof \subseteq dcof$. Consequently,

$$\begin{aligned} |\text{ind}(f, q, x_i)| &= |\text{cof}(f, q \cup \{x_i\}) \setminus subdcof| \\ &= |\text{cof}(f, q \cup \{x_i\})| - |subdcof| \\ &\geq |\text{cof}(f, q \cup \{x_i\})| - |dcof| \\ &\geq |\text{cof}(f, q \cup \{x_i\})| - 2 \cdot |\text{dep}(f, q, x_i)| \qquad (3.20) \end{aligned}$$

since $2 \cdot |\text{dep}(f, q, x_i)|$ is an upper bound for the number of direct cofactors in $x_i$ of functions in $\text{dep}(f, q, x_i)$. Now we have

$$\begin{aligned} &|\text{cof}(f, q \cup \{x_i\})| - |\text{dep}(f, q, x_i)| \\ &= |\text{cof}(f, q \cup \{x_i\})| - 2 \cdot |\text{dep}(f, q, x_i)| + |\text{dep}(f, q, x_i)| \\ &\leq |\text{ind}(f, q, x_i)| + |\text{dep}(f, q, x_i)| \qquad (3.21) \\ &= |\text{cof}(f, q)| \end{aligned}$$

and the result follows. Inequality (3.21) holds with Inequality (3.20). □

We are now able to prove the desired monotonicity of $h$.

PROPOSITION 5 *Let* $f\colon \mathbf{B}^n \to \mathbf{B}^m$ *be a Boolean multi-output function which essentially depends on every variable in* $X_n$. *The heuristic function* $h\colon 2^{X_n} \to \mathbb{N}$

$$h(q) = \max\left(|\mathrm{cof}(f, X_n \setminus q)| - \mathrm{roots}(X_n \setminus q), n - |q|\right)$$

*is* monotone *with respect to the* $A^*$*-algorithm outlined in Section 3.2.1.*

**Proof.** Like in the proof to Proposition 3, we first assert that in fact we have $h(X_n) = 0$ as required by definition of heuristic functions. Then we have to show again the validity of Inequality (3.13) in Section 3.2.3, i.e. we show

$$h(q) \leq |\mathrm{dep}(f, q, x_i)| + h(q \cup \{x_i\}). \tag{3.22}$$

Again, the proof is a careful case analysis of the possible outcomes for the maximum function.

**Case 1):**

$$\begin{aligned} |\mathrm{cof}(f, X_n \setminus q)| - \mathrm{roots}(X_n \setminus q) &> n - |q|, \\ |\mathrm{cof}(f, X_n \setminus q \cup \{x_i\})| - \mathrm{roots}(X_n \setminus q \cup \{x_i\}) &> n - |q \cup \{x_i\}| \end{aligned}$$

Inequality (3.22) follows directly with Lemma 3.24 and Lemma 3.23.

**Case 2):**

$$\begin{aligned} |\mathrm{cof}(f, X_n \setminus q)| - \mathrm{roots}(X_n \setminus q) &\leq n - |q|, \\ |\mathrm{cof}(f, X_n \setminus q \cup \{x_i\})| - \mathrm{roots}(X_n \setminus q \cup \{x_i\}) &\leq n - |q \cup \{x_i\}| \end{aligned}$$

We have to show

$$n - |q| \leq |\mathrm{dep}(f, q, x_i)| + n - |q \cup \{x_i\}|,$$

which has already been proven in Case 2) of Proposition 3 in Section 3.2.3.

**Case 3):**

$$\begin{aligned} |\mathrm{cof}(f, X_n \setminus q)| - \mathrm{roots}(X_n \setminus q) &> n - |q|, \\ |\mathrm{cof}(f, X_n \setminus q \cup \{x_i\})| - \mathrm{roots}(X_n \setminus q \cup \{x_i\}) &\leq n - |q \cup \{x_i\}| \end{aligned}$$

The proof is analogous to the one given for Case 3) of Proposition 3 in Section 3.2.3.

**Case 4):**

$$\begin{aligned} |\text{cof}(f, X_n \setminus q)| - \text{roots}(X_n \setminus q) &\leq n - |q|, \\ |\text{cof}(f, X_n \setminus q \cup \{x_i\})| - \text{roots}(X_n \setminus q \cup \{x_i\}) &> n - |q \cup \{x_i\}| \end{aligned}$$

The proof is analogous to the one given for Case 4) of Proposition 3 in Section 3.2.3. □

#### 3.2.5.2 Avoiding Transitions to Successor States

The top-down approach described in Section 3.2.4 uses a very efficient state expansion technique (see Section 3.2.4.5). By that technique, all but one variable shift needed to extend a state $q$ to a successor $q'$ (i.e. to build the transition $q \longrightarrow q'$) becomes obsolete: the only variable shift still left to perform *only once* the *first time* a successor state $q'$ is encountered is the one in line (10) in macro UPDATE-STATE-DATA-ASTAR-TD in Figure 3.22, triggered by the conditional statement in line (9). This shift prepares the computation of the heuristic function $h$. Without the expansion technique given in Section 3.2.4.5, much more variable shifts would be necessary: a successor state $q'$ with $|q'| = k$ is being visited up to $k$ times in the progress of one step of the minimization algorithm since $k$ distinct states $p \subset q$ with $|p| = k-1$ have the successor $q'$ in common. With the standard approach to state expansion, every new visit would force a new variable shift. With the suggested expansion technique, revisiting $q'$ does *not* involve any further variable shifts: $h(q')$ already has been computed the first time $q'$ was visited and hence, the variable shift of line (10) is not executed again.

This idea can be seen as an *acceleration* of the "early pruning" techniques of the algorithm JANUS presented in Section 3.1: assume a top-down approach. When revisiting successor states, JANUS avoids repeated variable shifts where they can be identified as unneeded. This is done using two lower bounds. A first lower bound lb_cost: $2^{X_n} \longrightarrow \mathbb{N}$ is used to decide whether revisiting a state $q$ still contributes to the computation of min_cost$(q, q)$, a quantity which corresponds to $g^*(q)$ in the $A^*$-based approach. If this is not the case, the algorithm does not visit $q$ again. A second bound used is lb_combined: $2^{X_n} \longrightarrow \mathbb{N}$;

$$\text{lb\_combined}(q) = h_\alpha(q) + h_\omega(X_n \setminus q) + 1 \tag{3.23}$$

where $h_\alpha$ denotes the heuristic function $h$ defined in Section 3.2.1 and $h_\omega$ denotes the heuristic function $h$ defined in Section 3.2.5. As usual, it excludes states from further examination. But now this already happens during the progress of each of the iteration steps. This is possible since lb_combined($q$), in contrast to lower bounds used in previous approaches, does not depend on the quantity min_cost($q, q$). Hence, it can already be computed before min_cost($q, q$) is determined. That way, state exclusion is done *earlier* than for the bounds suggested in former approaches. A state was excluded by these bounds at the end of each step at the earliest. As soon as a state is excluded, no further revisits take place.

While JANUS cannot *always* avoid the expensive variable shift caused by revisiting a state, the suggested bounds have been shown to be effective, see Section 3.1.4.

#### 3.2.5.3 Early Pruning

Since the proposed state expansion technique yields a large gain in run time, it would be desirable to transfer it to the bottom-up approach. Unfortunately, this seems impossible; in the BDD representing a state to expand, the required cofactors are not represented by a single BDD node. Moreover, the set of cofactors to consider changes with every variable by which a state is extended to a successor state. Hence it can be conjectured that an efficient graph traversing routine identifying and counting dependent cofactors is not available.

However, it is possible to use the "early pruning" techniques of JANUS in the bottom-up approach. Moreover, the method of delaying the computation of $h$ as described in Paragraph "Delayed State Insertion" in Section 3.2.4.1 can be integrated into the pruning techniques. Assume a bottom-up approach: a problem is that for a state $q$, lb_combined($q$) can not be used until $h_\omega(q) = h(q)$ is known (see Equation (3.23)). Therefore, in order to be able to exclude states right from the start, a weaker form of this lower bound, lb_combined_weak($q$), is used until lb_combined($q$) becomes available. Let $h_\alpha^{weak}\colon 2^{X_n} \to \mathbb{N}$, $h_\omega^{weak}\colon 2^{X_n} \to \mathbb{N}$;

$$\begin{aligned} h_\alpha^{weak}(q) &= \max\left(\text{label}(F, |q|+1), n-|q|\right), \\ h_\omega^{weak}(q) &= \max\left(\text{label}(F, |X_n \setminus q|+1) - \text{roots}(X_n \setminus q), n-|q|\right). \end{aligned}$$

These are weaker forms of the heuristic functions proposed before: basically, in the corresponding "strong" forms, a term $|\text{cof}(f, q)|$ for a state $q$ has been replaced by label($F, |q|+1$). By Lemma 3.14, these nodes of level $|q|+1$ represent cofactors with respect to the first $q$ variables of the ordering, i.e. they are a subset of cof($f, q$). Thus, we clearly have $h_\alpha^{weak}(q) \leq h_\alpha(q)$ and $h_\omega^{weak}(q) \leq h_\omega(q)$ for all states $q$. The weaker form of the lower bound lb_combined_weak: $2^{X_n} \to \mathbb{N}$ is now defined analogously to lb_combined($q$):

$$\text{lb\_combined\_weak}(q) = h_\alpha^{weak}(X_n \setminus q) + h_\omega^{weak}(q) + 1$$

A sketch of the resulting code for the bottom-up version of the $A^*$-based approach to exact BDD minimization is given in Figure 3.25. The algorithm is quite similar to the top-down approach in Figure 3.19, with the following differences: the top-down approach uses the state expansion technique described in Section 3.2.4.5 (see line (17) in Figure 3.19). Unfortunately, this is not possible for the bottom-up approach as has been explained before in this section. As a consequence, the bottom-up approach needs to perform variable shifts to construct the successor states during a state expansion (see line (24) in Figure 3.25). To be able to restrict their number with the already known techniques for state exclusion (as described above in this section and given in Figure 3.26), in line (26) the macro UPDATE-STATE-DATA-ASTAR-BU is used (see Figure 3.27 for the code of this macro).

In line (12) of Figure 3.27, the lower bound lb_combined is used if the heuristic function has already been computed, otherwise lb_combined_weak is used (line (16) in Figure 3.27).

### 3.2.6 Experimental Results

All experimental results have been carried out on a machine with an Athlon processor running at 1.4 GHz, with a main memory of 1.5 GByte and a run time limit of 20,000 CPU seconds. The $A^*$-based algorithm in its final stage of development is called $A^{stute}$. In the experiments also an earlier version of the algorithm has been included without the path and BDD reconstruction techniques described in Section 3.2.4.4: this algorithm is called $A^{stir}$. Instead of using path reconstruction, complete paths to states are stored with each state and instead of the BDD reconstruction technique given in Section 3.2.4.4, the technique presented in Section 3.1.3.2 is used. In the experiments $A^{stute}$ has been compared to the best known classical B&B method, called FizZ and to the approach JANUS. Both have been described in Section 3.1.

The implementation of the $A^*$-based algorithms $A^{stute}$ and $A^{stir}$ is based on the implementation of JANUS. All algorithms have been integrated in the CUDD package [Som02]. By this it is guaranteed that they run in the same system environment.

In a first series of experiments, a plain B&B-algorithm is stepwise turned into the $A^*$-based approach $A^{stute}$, thereby examining the effect of the new techniques incorporated into $A^*$ as explained in Section 3.2.4 in detail. In the first column of Table 3.3 the name of the function is given. Column *in* (*out*) gives the number of inputs (outputs) of a function. Column *opt* shows the number of BDD nodes needed for the minimal representation. In column *B&B*, the run time in CPU seconds for a B&B algorithm for the best plain B&B approach called FizZ (see Section 3.1)

```
compute_optimal_ordering(BDD F, int n)
  proc
    INIT-ASTAR
    while 1 do   /* until doomsday */
      determine q ∈ OPEN with minimal
      g[hash(q)] + h[hash(q)];
      CHECK-EARLY-TERMINATION
      if q = X_n then
        reconstruct ordering as the sequence of the path
        to q;
        doomsday:=1;
      end–if
      if doomsday then
        exit while loop;
      end–if
      reconstruct an appropriate ordering π ∈ Π(X_n \ q)
      with path reconstruction;
      establish π on F;
      upper_bound := update_upper_bound();
      for each x_i ∈ X_n \ q do
        CHECK-SYMMETRY
        q' := q ∪ {x_i};
        if q' ∉ states then
          g[hash(q')] := ∞;
        end–if
        CHECK-EXCLUDED-ASTAR-BU
        shift x_i to level n − |q|;
        newcost := label(F, n − |q|) + g[hash(q)];
        UPDATE-STATE-DATA-ASTAR-BU
        UNDO-SHIFT
      end–for
    end–while
  end–proc
```

*Figure 3.25.* Bottom-up approach of $\boldsymbol{A}^{stute}$.

is given. Column *plain* $A^*$ states the run time for what essentially is the $A^*$-approach as introduced in Section 3.2.4, but *without* incorporating the B&B paradigm. The implementation here was based on JANUS (which is based on FizZ). Next, column $A^*$*/B&B* gives the run time for the combination of the two paradigms $A^*$ and B&B. The effect of

```
CHECK-EXCLUDED-ASTAR-BU
  macro
    if q' already excluded or lb_cost(q') ≥ g[hash(q')] then
      continue with for-loop;
    end–if
  end–macro
```

*Figure 3.26.* Checking for exclusion (early pruning).

```
UPDATE-STATE-DATA-ASTAR-BU
  macro
    if q' ∉ states or newcost < g[hash(q')] then
      g[hash(q')] := newcost;
      p(q') = x_i;
      upper_bound := update_upper_bound();
      if q' ∉ states then
        states[hash(q')] := q';
      end–if
      if h[hash(q')] not yet computed and
      g[hash(q')] + n − |q'| + 1 ≤ upper_bound then
        h[hash(q')] := compute_lower_bound(q');
        if lb_combined(q') > upper_bound then
          exclude q';
        end–if
      else
        if lb_combined_weak(q') > upper_bound then
          exclude q';
        end–if
      end–if
      if h[hash(q')] computed and
      g[hash(q')] + h[hash(q')] + 1 ≤ upper_bound then
        if OPEN too crowded, resize it and
        garbage-collect outdated entries;
        insert q' into OPEN;
      end–if
    end–if
  end–macro
```

*Figure 3.27.* Updating the state data in the $A^*$-based bottom-up approach.

*Table 3.3.* Effect of the techniques in $\boldsymbol{A}^{stute}$

| name | in | out | opt | B&B | plain $A^*$ | $A^*$/B&B | expand | $A^{stute}$ |
|---|---|---|---|---|---|---|---|---|
| cc | 21 | 20 | 46 | 117s | 65.7s | 59.6s | 54.3s | 51.1s |
| cm150a | 21 | 1 | 33 | 610s | 241s | 208s | 124s | 119s |
| comp | 32 | 3 | 95 | 5606s | 3859s | 3533s | 1702s | 1477s |
| cordic | 23 | 2 | 42 | 3.05s | 2.65s | 2.31s | 1.43s | 1.29s |
| cps | 24 | 102 | 971 | 4396s | 2406s | 2380s | 1590s | 1380s |
| i1 | 25 | 16 | 36 | 29.4s | 21.67s | 18.04s | 19.93s | 18.0s |
| lal | 26 | 19 | 67 | 677s | 436s | 402s | 271s | 242s |
| mux | 21 | 1 | 33 | 610s | 240s | 208s | 124s | 119s |
| pcle | 19 | 9 | 42 | 9.02s | 5.96s | 5.37s | 4.18s | 3.75s |
| pm1 | 16 | 13 | 40 | 0.55s | 0.62s | 0.57s | 0.53s | 0.54s |
| s208.1 | 18 | 9 | 41 | 8.44s | 7.23s | 6.52s | 4.15s | 3.72s |
| s298 | 17 | 20 | 74 | 13.46s | 8.65s | 8.43s | 6.38s | 6.00s |
| s344 | 24 | 26 | 104 | 1446s | 955s | 844s | 630s | 527s |
| s349 | 24 | 26 | 104 | 1447s | 959s | 844s | 631s | 527s |
| s382 | 24 | 27 | 119 | 802s | 503s | 469s | 460s | 390s |
| s400 | 24 | 27 | 119 | 802s | 501s | 469s | 460s | 390s |
| s444 | 24 | 27 | 119 | 779s | 484s | 461s | 395s | 324s |
| s510 | 25 | 13 | 146 | 13224s | 6985s | 5791s | 3787s | 3308s |
| s526 | 24 | 27 | 113 | 1196s | 631s | 600s | 445s | 391s |
| s820 | 23 | 24 | 220 | 2034s | 1257s | 1194s | 847s | 733s |
| s832 | 23 | 24 | 220 | 2076s | 1249s | 1193s | 847s | 730s |
| sct | 19 | 15 | 48 | 8.62s | 7.36s | 6.74s | 4.22s | 3.83s |
| tcon | 17 | 16 | 25 | 0.52s | 0.73s | 0.57s | 0.57s | 0.52s |
| ttt2 | 24 | 21 | 107 | 950s | 610s | 574s | 416s | 362s |
| vda | 17 | 39 | 478 | 65.4s | 34.8s | 32.5s | 31.9s | 25.8s |

using the state expansion technique of Section 3.2.4.5, avoiding many variable shifts, can be seen in column *expand*. Finally, the last column $A^{stute}$ gives the run times of the fine-tuned algorithm in its final stage of development, using the efficient method for lower bound computation described in Section 3.1.2.

As expected, the plain B&B method yields the largest run times. Using the introduced $A^*$-paradigm with the path reconstruction technique instead can reduce run time to much less than half the one needed for B&B (e.g., see *cm150a, mux*). The improvement is robust, yielding consistent results. Combining B&B and $A^*$, thus incorporating the techniques of delayed state insertion and of early termination as explained in Section 3.2.4.1, for almost every test-case yields a further, significant improvement. If the state expansion technique of Section 3.2.4.5 is used in addition, the reductions in run time can be high for some of the larger test-cases (e.g., see *comp*) and is still significant for most of the other, smaller test-cases. Due to incorporating some of the fine-tuning of JANUS, the final algorithm $A^{stute}$ shows a further improvement on all but the smallest test-cases. These reductions in run time are sometimes even higher than those obtained by using the improved state expansion technique (see *s382*, *s400*). Comparing the final algorithm $A^{stute}$ to the best known "classical" B&B-method which is following the standard approach using only one lower bound, called FizZ (see Section 3.1), a reduction in run time of up to 80.5% has been obtained (see *cm150a*, *mux*). On average, the speed-up is 69.8%. In this, using the optimized $A^*$-paradigm instead of a plain B&B-paradigm yields a consistent and robust reduction in run time of 70-80% (a speed-up of three up to five times faster), clearly demonstrating the efficiency of the approach.

For arithmetic functions, i.e. adders and multipliers, the reduction in run time by going from B&B to $A^*$ can be even higher, as the results of the next test-suite demonstrate. Moreover, the fact that arithmetic functions are easily scalable by the number of input variables $n$, allows to analyze the development of the speed-up with $n$.

In a second series of experiments arithmetic functions are considered, i.e. adders and multipliers (see Tables 3.4, and 3.5). The object of investigation here was a comparison of the run times for the first to the last algorithm in the series of methods ranging from plain B&B to the fine-tuned $A^{stute}$-approach, as studied in detail in the first series of experiments. First a plain B&B method (i.e. FizZ) and the top-down approach of $A^{stute}$, in the following denoted $A^{stute}\downarrow$, has been applied both to adders and multipliers, see Table 3.4. In columns time and space the run time in CPU seconds and the space requirement in MByte for the two approaches are given, respectively. In the case of adder functions, for the set of test-cases, the results show a tremendous speed-up of up to one order of magnitude when using $A^{stute}\downarrow$ instead of plain B&B, i.e. FizZ has been outperformed significantly. Moreover, both for adder and multiplier functions the speed-up appears to grow linear with $n$, the number of input variables of the arithmetic circuit. This can be seen in Figure 3.28 and Figure 3.29.

In Table 3.5 the run times for the bottom-up version of $A^{stute}$, in the following denoted $A^{stute}\uparrow$ when applied to multipliers are compared to the previous results with $A^{stute}\downarrow$. The top-down approach failed to minimize *mult9* within the given time-limit of 20,000 seconds. In Section

*Table 3.4.* Minimizing arithmetic functions with B&B and $\boldsymbol{A}^{stute}\downarrow$

| name | in | out | opt | B&B | | $A^{stute}\downarrow$ | |
|---|---|---|---|---|---|---|---|
| | | | | time | space | time | space |
| adder8 | 16 | 8 | 36 | 0.11s | <1 | 0.39s | <1 |
| adder12 | 24 | 12 | 56 | 1.73s | <1 | 0.92s | <1 |
| adder16 | 32 | 16 | 76 | 15.0s0 | <1 | 4.02s | 2 |
| adder20 | 40 | 20 | 96 | 88.4s | 4 | 17.4s | 6 |
| adder24 | 48 | 24 | 116 | 426s | 8 | 65.3s | 19 |
| adder28 | 56 | 28 | 136 | 1842s | 34 | 214s | 38 |
| adder32 | 64 | 32 | 156 | 8075s | 76 | 645s | 142 |
| mult2 | 4 | 2 | 12 | 0.01s | <1 | 0.34s | <1 |
| mult3 | 6 | 3 | 41 | 0.02s | <1 | 0.38s | <1 |
| mult4 | 8 | 4 | 135 | 0.06s | <1 | 0.37s | <1 |
| mult5 | 10 | 5 | 388 | 0.66s | <1 | 0.62s | <1 |
| mult6 | 12 | 6 | 1098 | 9.21s | <1 | 4.36s | 3 |
| mult7 | 14 | 7 | 3082 | 218.2 | <1 | 85.9s | 8 |
| mult8 | 16 | 8 | 8658 | 13580s | 3 | 3878s | 25 |

*Table 3.5.* Minimizing multiplier functions top-down and bottom-up

| name | in | out | opt | $A^{stute}\downarrow$ | | $A^{stute}\uparrow$ | |
|---|---|---|---|---|---|---|---|
| | | | | time | space | time | space |
| mult2 | 4 | 2 | 12 | 0.34s | <1M | 0.32s | <1M |
| mult3 | 6 | 3 | 41 | 0.38s | <1M | 0.34s | <1M |
| mult4 | 8 | 4 | 135 | 0.37s | <1M | 0.38s | <1M |
| mult5 | 10 | 5 | 388 | 0.62s | <1M | 0.61s | <1M |
| mult6 | 12 | 6 | 1098 | 4.36s | 3M | 1.88s | 3M |
| mult7 | 14 | 7 | 3082 | 85.9s | 8M | 18.45s | 8M |
| mult8 | 16 | 8 | 8658 | 3878s | 25M | 422s | 23M |
| mult9 | 18 | 9 | 24326 | – | – | 7340s | 71M |

3.1.4 it has been demonstrated that, for multipliers, the run time of the bottom-up approach of JANUS, i.e. JANUS$\uparrow$, is significantly shorter than that of JANUS$\downarrow$, achieving speed-ups of one order of magnitude. Similar results have been observed comparing the bottom-up version

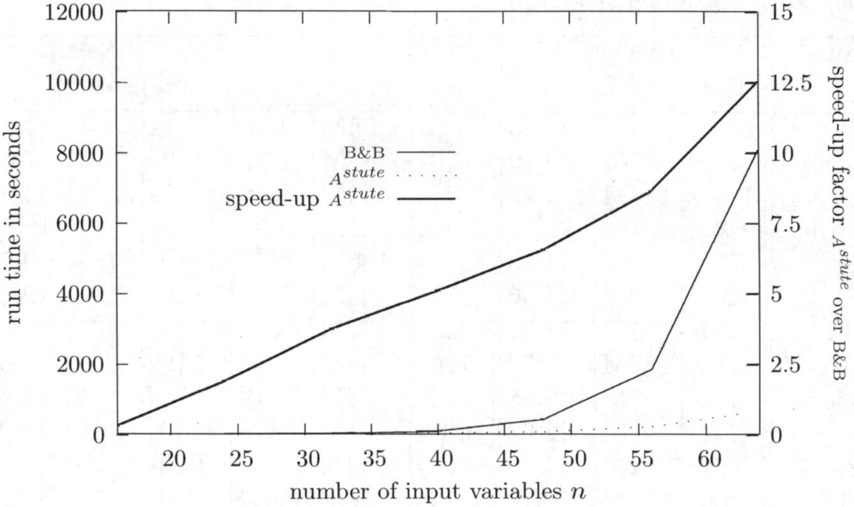


*Figure 3.28.* Run times of B&B and $\boldsymbol{A}^{stute}$ for adder functions.

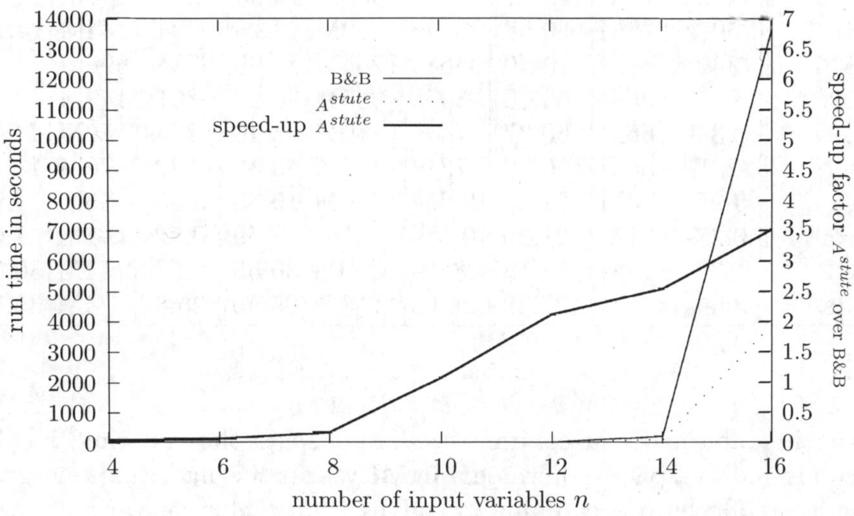


*Figure 3.29.* Run times of B&B and $\boldsymbol{A}^{stute}$ for multiplier functions.

$A^{stute}\downarrow$ with the top-down version $A^{stute}\uparrow$, also see Figure 3.30. The speed-up even grows faster than linear with $n$ in this case.

In a next series of experiments all algorithms have been applied to the set of benchmark circuits from LGSynth93 [Col93]. The results are

*Table 3.6.* Comparison of JANUS, $\boldsymbol{A}^{stir}$ and $\boldsymbol{A}^{stute}$

| name | JANUS | | | $A^{stir}$ | | | $A^{stute}$ | | |
|---|---|---|---|---|---|---|---|---|---|
| | time | space | a.s. | time | space | a.s. | time | space | a.s. |
| cc | 84.9s | 34M | 47.43 | 51.6s | 85M | 14.48 | 51.1s | 36M | 14.60 |
| cm150a | 311s | 35M | 49.79 | 120s | 88M | 7.22 | 119s | 39M | 7.63 |
| comp | 3900s | 124M | 126.22 | 1894s | 649M | 29.86 | 1477s | 288M | 33.96 |
| cordic | 1.82s | < 1M | 25.43 | 1.35s | < 1M | 18.82 | 1.29s | < 1M | 19.17 |
| cps | 2751s | 58M | 56.13 | 1373s | 186M | 39.97 | 1380s | 78M | 40.05 |
| i1 | 18.8s | 10M | 53.36 | 17.2s | 25M | 12.87 | 18.0s | 11M | 12.98 |
| lal | 504s | 75M | 74.72 | 243s | 378M | 25.01 | 242s | 145M | 25.50 |
| mux | 311s | 35M | 74.72 | 120s | 88M | 7.24 | 119s | 39M | 7.68 |
| pcle | 5.18s | 3M | 22.88 | 3.61s | 6M | 12.79 | 3.75s | 3M | 13.60 |
| pm1 | 0.34s | < 1M | 22.62 | 0.61s | < 1M | 10.25 | 0.54s | < 1M | 11.23 |
| s208.1 | 5.62s | 2M | 24.75 | 3.84s | 4M | 17.79 | 3.72s | 3M | 17.10 |
| s298 | 9.06s | 2M | 29.13 | 6.17s | 6M | 17.52 | 6.00s | 3M | 17.56 |
| s344 | 950s | 105M | 71.44 | 537s | 363M | 28.74 | 527s | 146M | 29.10 |
| s349 | 950s | 105M | 71.44 | 537s | 363M | 28.74 | 527s | 146M | 29.10 |
| s382 | 461s | 71M | 49.41 | 313s | 359M | 14.06 | 390s | 143M | 14.07 |
| s400 | 456s | 71M | 49.41 | 312s | 359M | 14.06 | 390s | 143M | 14.07 |
| s444 | 508s | 78M | 50.58 | 314s | 362M | 14.25 | 324s | 144M | 14.16 |
| s510 | 7342s | 386M | 75.40 | 3568s | 1264M | 25.05 | 3308s | 596M | 24.55 |
| s526 | 924s | 105M | 63.85 | 365s | 365M | 30.26 | 391s | 149M | 30.25 |
| s820 | 1235s | 56M | 60.37 | 734s | 180M | 43.46 | 733s | 76M | 43.45 |
| s832 | 1288s | 56M | 60.09 | 731s | 181M | 43.46 | 730s | 76M | 43.41 |
| sct | 5.97s | 3M | 27.08 | 3.86s | 6M | 19.85 | 3.83s | 3M | 19.90 |
| tcon | 0.28s | < 1M | 21.29 | 0.55s | 1M | 7.69 | 0.52s | < 1M | 7.69 |
| ttt2 | 578s | 78M | 55.15 | 345s | 360M | 30.55 | 362s | 144M | 30.59 |
| vda | 34.4s | 3M | 27.30 | 26.0s | 6M | 35.50 | 25.8s | 4M | 35.54 |

given in Table 3.6. In columns *time* and *space* the run time in CPU seconds and the space requirement in MByte for the algorithms are given. For every algorithm, column *a.s.* gives the number of variable swaps needed on average to set the variable ordering for the BDD representing the next state to "process" (i.e. to extend its corresponding ordering in the case of JANUS or to expand it to its successor states in the case of the $A^*$-based approaches).

As the results in Table 3.6 show, algorithm $A^{stute}$ has a reduced memory requirement when compared to the earlier version $A^{stir}$ by up to

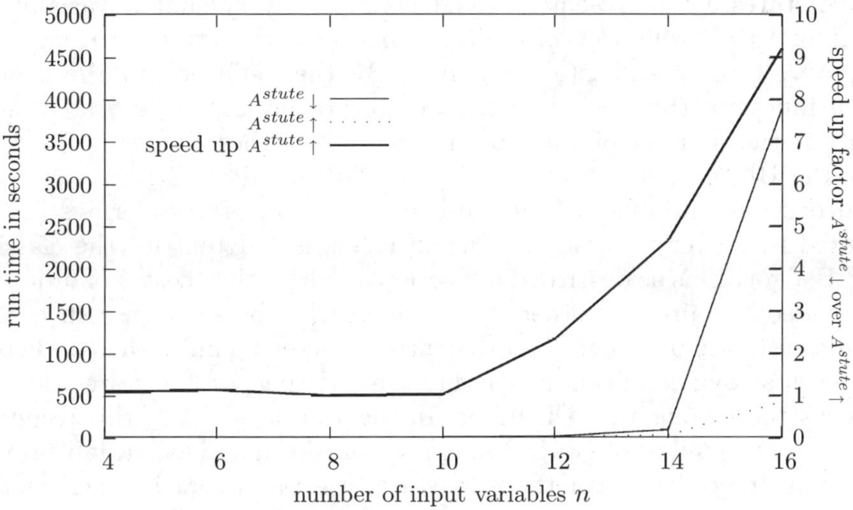


*Figure 3.30.* Run times of $\boldsymbol{A}^{stute}\downarrow$ and $\boldsymbol{A}^{stute}\uparrow$ for multiplier functions.

61.6% (e.g., see *lal*), on average the reduction is 57.5%. The technique to avoid the storing of complete orderings is a significant, non-heuristic and thus robust improvement successfully addressing the problem of higher memory consumption of $A^*$-based approaches. In this, with the memory configuration of todays computer systems, the $A^*$-based approach remains practical. The BDD reconstruction technique of $A^{stute}$ reduces run time by up to 22.0% compared to $A^{stute}$ (e.g., see *comp*). This method successfully avoids BDD explosions and achieves a significant speed-up. This holds especially for the largest examples which are more sensitive to variable ordering. Algorithm $A^{stute}$ is much faster than algorithm JANUS. In comparison to JANUS, reductions in run time of more than 60% can be obtained (e.g., see *comp*, *mux*, *cm150a*). On average a gain of 50.8% has been achieved. One reason for this large gain is the reduction in the number of state expansions which can be up to 40.9%: e.g. for *s526*, $A^{stute}$ expands only 852179 states whereas JANUS extends 1442625 variable orderings by one variable (which is the equivalent to a state expansion in $A^*$). As can be expected from search theory, the best-first strategy can yield large reductions in the number of state expansions especially for the "harder" instances (more complex circuits) which directly transfers to a reduction in run time.

In the experiments, a second reason for the high speed-up of $A^{stute}$ compared to JANUS has been identified: typically, many states with the same $\varphi$-value are reconstructed for expansion one after the other.

Such states are represented by BDDs with a very similar variable ordering: to see this, let $q, q' \subseteq X_n$ be such two states, i.e. we have $|q| = |q'| =: k$ and $\varphi(q) = \varphi(q')$. By the definition of the evaluation function, the sum of a) the number of nodes in the first $k$ levels and b) the number of cofactors referenced by these nodes is the same for the BDDs representing $q$ and $q'$. This condition strongly restricts the possible variable orderings for $q$ and $q'$, i.e. the orderings are required to be very "similar". The situation is different for the classical B&B algorithms as described in Section 3.1: here the "costs" of orderings extended one after the other are *not* required to be equal (it suffices that these costs are both less than the current upper bound). Hence, there is no such strong condition restricting the orderings of two states that are reconstructed one after the other. In the approach $A^{stute}$, this results in an average number of swaps needed to transform a given variable ordering into the ordering for the next state to expand that is much smaller than in previous approaches, yielding significant smaller run times for BDD reconstruction: in the experiments, this average number of variable swaps is reduced by up to 84.7% in comparison to JANUS. E.g. for *cm150a* the average number of swaps needed by JANUS is 49.79, but only 7.63 for $A^{stute}$. On average, a reduction in the number of variable swaps of 56.0% has been obtained.

## 3.3 Summary

In this chapter, classical and recently published methods for exact BDD minimization have been presented. For this purpose we started with an observation of [FS90] regarding the invariance of level sizes with respect to certain restricted variable movements. The first method yielding a significant improvement over a naive brute-force search was based on this observation.

Next, approaches based on B&B to prune the search space have been described. The most recent one is called JANUS [EGD03b] and applies an extended B&B-technique where more than one lower bound is used in parallel. This technique enables us to often avoid repeatedly visiting states. The lower bounds have been derived by generalization of a lower bound known from VLSI design. Moreover, a faster method of lower bound computation as well as an efficient method of partial BDD reconstruction, which avoids many time consuming variable shifts have been given.

The presentation of JANUS was finished by giving experimental results that clearly demonstrate the efficiency of both the presented top-down and bottom-up approach. A comparison to the best previous minimization algorithm shows that run time can be reduced by up to 49%.

The bottom-up approach of JANUS achieves speed-ups of two orders of magnitude compared to the best previous bottom-up approach.

The latest development in exact BDD minimization is the shift to a new paradigm, the $A^*$-algorithm. This is a search technique frequently used in AI.

Recently, an $A^*$-based approach called $A^{stute}$ has been suggested [EGD05]. In contrast to the classical approaches, search is reduced to the computation of an optimal path in a state space. By a best-first ordering of states, larger parts of the search space can be pruned than in all approaches presented so far. Techniques to *combine* the $A^*$-algorithm with classical B&B methods have been described, resulting in a further pruning of states, reduction of memory requirement and less computations during the algorithm run. A technique to reconstruct paths rather than persistently storing them was shown to largely reduce memory requirement of $A^*$-based approaches. In the application of exact BDD minimization, the reduction is almost 60% on average. A unique BDD reconstruction technique has been presented which is more effective than the best previously known technique. Moreover, the best-first ordering of states speeds up the time-consuming operation of BDD reconstruction significantly.

The presentation of the $A^*$-based approach to exact BDD minimization is completed by experimental results that clearly demonstrate its efficiency. A comparison to the most recent B&B-based method JANUS shows that run time can be reduced by more than 60%.

Search is a task which frequently occurs in different areas of VLSI CAD. Moreover, applying the paradigm of state space exploration will often help to reduce the complexity of typical search problems in advance. The $A^*$-algorithm, known from AI and well-founded on a thoroughly analyzed theory, offers very promising solutions to search tasks of time-limited nature.

Chapter 4

# HEURISTIC NODE MINIMIZATION

This chapter continues the presentation of classical and recent results achieved in the area of size-driven BDD optimization. The studies described in the previous chapter of this book have been aiming at exact minimization of BDDs. This was motivated by applications in logic synthesis where suboptimal solutions are a significant drawback, because they lead to increased chip areas. In the remaining part, the focus will be on *heuristic* methods. As has been pointed out earlier in Chapter 1, it is NP-complete to decide whether the number of nodes can be improved by variable reordering. Hence, it is an obvious idea to look for heuristic solutions, regardless of the fact that optimal solutions then cannot be guaranteed.

For many applications of BDDs optimal solutions are not needed, e.g. formal verification, model checking and symbolic state space representation. Using heuristic methods in these fields of application avoids the high run times of exact methods.

For this, in the past many heuristic approaches have been proposed that roughly can be classified as follows. The first type of heuristics tries to determine an appropriate variable ordering in advance, i.e. prior to BDD construction. These heuristics make use of the knowledge available about the respective type of problem, for example the structural information for a given circuit can be exploited, e.g. see [FOH93]. Heuristics of this type can be fast. However, they might fail to yield good orderings in intermediate steps of symbolic algorithms because the function represented at these steps is different from the one targeted with the heuristic. An example is the symbolic representation of a state transition relation with a BDD for its characteristic function. Typically, the represented set of transitions is increased by new transitions every now and again during the construction of the set. As the function represented changes with every step, the quality of the initial ordering is often degraded. Another example is the successive creation of a BDD representing a circuit given as a netlist: again the function evolves and changes with every new BDD operation during the construction.

The second type of heuristics tackles this problem by *dynamic reordering*. Here, no "a-priori"-knowledge about the particular application is necessary. Instead, a better ordering is determined dynamically by the system in situations where the application gets low on memory. This typically happens in a fully automated manner and does not involve any inputs or actions from the user of such a software system.

The idea of dynamic reordering has been raised in [FMK91] and [ISY91]. Both works suggest a *windows permutation algorithm* to perform the reordering: basically, an exhaustive search for better orderings is performed, but only within windows of constant size $k$. In [Rud93], Rudell showed the limitations of this approach and suggested to use the *sifting algorithm* instead. This algorithm has been presented in detail in Section 2.4.7.2. Both approaches, the window permutation and the sifting algorithm, were based on the efficient exchange of adjacent variables (see Section 2.4.7.1).

However, as a BDD-based system may invoke dynamic reordering very often during the progress of BDD construction, run time was still an important issue. In many cases dynamic reordering seemed to be too time-consuming to replace the methods which determine the ordering in advance. Consequently, the demand for faster, automated solutions was high.

For this reason in [MS97] an algorithm has been proposed how to partition the search space by grouping variables to improve sifting run times. However, this method is known to be strongly dependent on the initial ordering. Another approach suggested search space partitioning by means of sampling [SM98], but the quality of the results varies widely depending on the choice of candidate variables, i.e. the choice of the sample.

In the struggle for better solutions, very promising results have been obtained by methods to prune the search space with the use of lower bounds on BDD sizes during sifting, called *lb-sifting* and *elb-sifting* [DGS01, ED05]. These lower bounds state minimum sizes for certain orderings that will be considered in the following steps of the sifting algorithm. They are used to limit the range of possible moves for each variable, and large reductions in run time are achieved by focusing only on those parts of the search space where improvements are possible.

Note that this increases efficiency without changing the quality of the results, i.e. the sizes of the resulting BDDs are preserved.

This chapter is focused on the method of sifting with lower bounds. Classical and recently published lower bounds on BDD sizes, applicable in dynamic reordering, are presented.

Lower bounds that lately have been suggested are derived by adapting more general lower bounds which have been presented in Chapter 3. They were originally intended for the use in exact BDD minimization and now they are transferred to the context of dynamic reordering.

First, deeper insight is gained by looking at the theory of the lower bounds. Examples are given which show that the lower bounds recently suggested behave "orthogonally" to the classical lower bounds, i.e. they are effective in situations where the previous ones are not and vice versa. This leads to a better understanding of the different impact of lower bounds on the efficiency of the sifting algorithm.

Since computation of the bounds is expensive, they are restricted to more efficient forms, following the constraints in practice. This compromises between computational complexity and pruning power, i.e. between run time and quality of the lower bounds.

Finally, a *combination* of classical and the recently published lower bounds is introduced, which fuses their capabilities to prune the search space in different situations. This yields a final lower bound, which is then incorporated into the sifting algorithm. Experiments show the significance of the obtained improvement.

## 4.1 Efficient Dynamic Minimization

A very effective method to reduce the number of variable swaps needed in sifting is the use of lower bounds on future BDD sizes. The size obtained by further movement of the considered variable cannot fall below the size stated by these lower bounds.

The idea is to stop moving the variable in the current direction (downward or upward) as early as possible. We can stop moving if the BDD size stated by the lower bound already exceeds the smallest BDD size recorded so far. No further improvement is possible, i.e. no better position for the considered variable can be found. Hence we can continue with the next variable without changing the results yielded by the method. In [DGS01] this idea of lower bound sifting (lb-sifting) has been introduced, together with effective lower bounds. The lower bounds have been derived from upper bounds on BDD sizes resulting from variable movements, as can be found in [BLW96]. Several techniques to derive the lower bounds have been used, e.g. inversion of equalities and specialization of general cases.

The following will be needed when considering lower bounds in dynamic reordering.

DEFINITION 4.1 *Let* $f = (f_i^n)_{1\leq i\leq m}$ *be a Boolean multi-output function. Two variables* $x_j, x_k \in X_n$ *are said to be* non-interacting *iff*

$$\nexists_{1\leq i\leq m}\colon\ x_j, x_k \in \text{support}(f_i).$$

*Otherwise, the variables are said to* interact.

In case of adjacency of two variables in the ordering of a BDD, the property of non-interaction of two variables is a sufficient condition for a swap of them being trivial, i.e. a constant operation. Indeed, if two variables do not interact, there is no arc connecting the two layers. Therefore, no edges must be redirected, no nodes vanish, etc. Hence the swap can be performed by exchanging two entries in a table mapping variable subscripts to the variables, i.e. in constant time. Interactivity of two variables is a necessary, but not a sufficient condition for non-triviality of the swap step (see [PS95, Som02] and also Section 2.4.7.1).

We denote the set of variables in $X_n$ interacting with a given variable $x_i \in X_n$ with $\mathcal{I}_{i,n}$. Also note that every variable interacts with itself, i.e. $x_i \in \mathcal{I}_{i,n}$ for every variable $x_i \in X_n$.

The classical lower bounds as supposed by [DGS01] are stated in the following result.

THEOREM 4.2 *Let* $F$ *be a BDD over* $X_n$*, for which we assume the natural variable ordering* $\pi$ *with* $\pi(i) = x_i$ $(1 \leq i \leq n)$*. Let* $|F_j{}'|$ *denote the size of the BDD after moving variable* $x_i$ *to position* $j$*. When moving down a variable* $x_i \in X_n$*, as a lower bound on the size of the resulting BDD* $F'$ *we have*

$$
\begin{aligned}
&\text{lb}^{\downarrow}(F, x_i)\\
&\quad= \min_{j=i+1,\ldots,n} |F_j{}'|\\
&\quad\geq \min_{j=i+1,\ldots,n} \{\text{label}(F, X_{i-1}^1) + \max\{\text{label}(F, X_j^{i+1} \setminus \mathcal{I}_{i,n}) + 1\\
&\qquad\qquad + \tfrac{1}{2} \cdot \text{label}(F, X_j^{i+1} \cap \mathcal{I}_{i,n}), \text{label}(F, x_i)\}\\
&\qquad\qquad + \text{label}(F, X_n^{j+1})\}\\
&\quad= \text{label}(F, X_{i-1}^1) + \max\{\text{label}(F, X_n^{i+1} \setminus \mathcal{I}_{i,n}) + 1\\
&\qquad + \frac{1}{2} \cdot \text{label}(F, X_n^{i+1} \cap \mathcal{I}_{i,n}), \text{label}(F, x_i)\},
\end{aligned}
$$

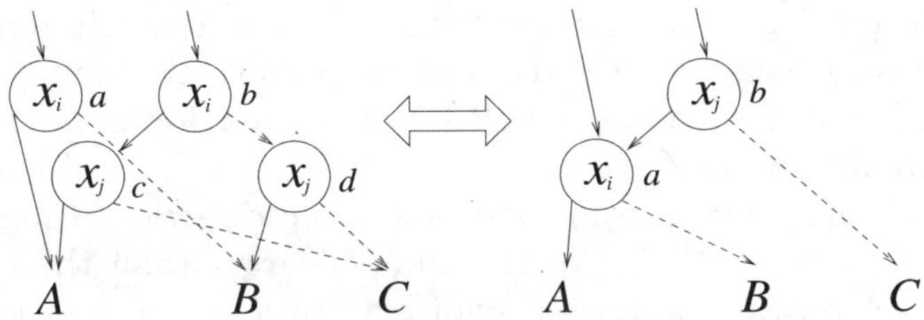


*Figure 4.1.* Result of the swap of neighbored variables.

*and when moving up, the lower bound is given as*

$$\begin{aligned}
\mathrm{lb}^{\uparrow}(F, x_i) &= \min_{j=1,\ldots,i-1} |F_j{}'| \\
&\geq \min_{j=1,\ldots,i-1} \{ \mathrm{label}(F, X^1_{j-1}) + \mathrm{label}(F, X^j_{i-1} \setminus \mathcal{I}_{i,n}) \\
&\qquad + \left| X^j_{i-1} \cap \mathcal{I}_{i,n} \right| + \frac{\mathrm{label}(F, x_i)}{2^{|X^j_{i-1} \cap \mathcal{I}_{i,n}|}} \\
&\qquad + \mathrm{label}(F, X^{i+1}_n) \} \\
&= \mathrm{label}(F, X^1_{i-1} \setminus \mathcal{I}_{i,n}) + \left| X^1_{i-1} \cap \mathcal{I}_{i,n} \right| + \frac{\mathrm{label}(F, x_i)}{2^{|X^1_{i-1} \cap \mathcal{I}_{i,n}|}} \\
&\qquad + \mathrm{label}(F, X^{i+1}_n).
\end{aligned}$$

Next, a summary of the basic ideas behind these bounds is given here, full proofs can be found in [DGS01].

A first (trivial) lower bound (already used in the BDD package CUDD [Som02] is the following: Assume variable $x_i$ is situated at level $i$. Moving $x_i$ down to level $j$, all nodes above level $i$ (i.e. in levels $1, \ldots, i-1$) and all nodes below level $j$ (i.e. in levels $j+1, \ldots n$) do not change due to the only local impact of the variable swap on the BDD structure (see Section 2.4.7.1). An analogous statement holds while moving $x_i$ up to level $j$. Furthermore, in both cases the following nodes also do not change: these are the nodes between levels $i$ and $j$ whose variable does not interact with $x_i$. Thus, the sum of these node numbers constitutes a first lower bound, counting those nodes, which are not affected by the swap. A more sophisticated idea is illustrated in Figure 4.1: it can be proven that the number of nodes in the levels affected by a swap can be at most reduced by a factor of two (for a proof see [DGS01]). The terms of the form $\frac{\mathrm{label}(F,*)}{2^*}$ occurring in the lower bounds given above are derived by (repeated) use of this argument.

Another idea used in the first lower bound for moving down a variable $x_i$ is the following: let the above BDD $F$ represent a Boolean function $f$. Assuming $x_i$ is situated at level $i$, the nodes labeled $x_i$ represent cofactors of $f$ with respect to the first $i-1$ variables of the ordering. This set of cofactors must still be represented in the diagram after moving $x_i$ downwards since otherwise the diagram would not represent $f$ anymore. Hence, the number of nodes labeled $x_i$ also is a lower bound on the number of nodes in levels $i, \ldots, n$.

During sifting, when moving into a specific direction, the according lower bound is used. If the lower bound is larger than the best BDD size found before, a movement cannot lead to a better position for the variable.

For $I \subseteq X_n$, the terms label$(F, I)$ can be computed very efficiently during sifting since the level sizes are kept in dedicated variables by modern BDD packages, e.g. see [Som02]. The question, if a variable interacts with $x_i$ also can be decided efficiently (i.e., in constant time) with a pre-computed "interaction matrix" giving the required information for every pair of variables in question (see [PS95]).

In [DGS01] it has been reported that using these lower bounds during sifting reduces run time by up to 70%.

## 4.2 Improved Lower Bounds for Dynamic Reordering

In this section lower bounds recently published for use in dynamic reordering are given. They are derived by adapting lower bounds which previously have been suggested to speed-up exact BDD minimization (see the approaches presented in Chapter 3) to the context of the sifting algorithm. Next, preliminary results preparing the proof of an improved lower bound for dynamic reordering are following.

LEMMA 4.3 *Let $F = (\pi, \ldots, O)$ be a BDD over $X_n$, for which we assume the natural variable ordering $\pi$ with $\pi(i) = x_i$ $(1 \leq i \leq n)$. Then, for all $0 \leq k \leq n$, we have*[1]

$$O = \mathcal{K}(F, 0) \subseteq \text{nodes}(F, X_k^1) \cup \mathcal{K}(F, k).$$

**Proof.** This can be shown by Lemma 3.17 which states the basic set inclusion used for the proof of the monotonicity of the heuristic function used in the $A^*$-approach to exact BDD minimization, described in Section 3.2.

Repeated application of Lemma 3.17 to the term $\mathcal{K}(F, 0)$, together with Equation (3.6) and Lemma 3.14, yields the required result.

[1]The definition of $\mathcal{K}(F, k)$ has been introduced in Definition 3.9.

In detail: by repeatedly applying Lemma 3.17 we obtain

$$\begin{aligned}
\mathrm{cof}(f,\emptyset) &\subseteq \mathrm{dep}(f,\emptyset,x_1) \cup \mathrm{cof}(f,\{x_1\}) \\
&\subseteq \mathrm{dep}(f,\emptyset,x_1) \cup \mathrm{dep}(f,\{x_1\},x_2) \cup \mathrm{cof}(f,\{x_1,x_2\}) \\
&\dots \\
&\subseteq \Big(\cup_{1\le k\le j}\mathrm{dep}(f,X^1_{k-1},x_k)\Big) \cup \mathrm{cof}(f,X^1_j),
\end{aligned}$$

which, applying Lemma 3.14 to the term $\cup_{1\le k\le j}\mathrm{dep}(f,X^1_{k-1},x_k)$ and applying Equation (3.6) to the terms $\mathrm{cof}(f,\emptyset)$, and $\mathrm{cof}(f,X^1_j)$, yields the required result. □

The following result follows from Lemma 4.3 and prepares the proof of a theorem stating an improved lower bound for dynamic reordering.

COROLLARY 4.4 *Let $F = (\pi,\dots,O)$ be a BDD over $X_n$, for which we assume the natural variable ordering $\pi$ with $\pi(i) = x_i$ $(1 \le i \le n)$. Let $1 \le i \le n$. Then, for all $1 \le j < i$, we have*

$$\begin{aligned}
&O \cap \mathrm{nodes}(F, X^1_i \cap \mathcal{I}_{i,n}) \\
&\quad\subseteq \mathrm{nodes}(F, X^1_{j-1} \cap \mathcal{I}_{i,n}) \cup \Big(\mathcal{K}(F,j-1) \cap \mathrm{nodes}(F, X^j_i \cap \mathcal{I}_{i,n})\Big).
\end{aligned}$$

**Proof.** Since $j \ge 1$, Lemma 4.3 yields

$$O \subseteq \mathrm{nodes}(F, X^1_{j-1}) \cup \mathcal{K}(F,j-1).$$

By intersection of both sides of this inclusion with $\mathrm{nodes}(F, X^1_i \cap \mathcal{I}_{i,n})$ we derive

$$\begin{aligned}
&O \cap \mathrm{nodes}(F, X^1_i \cap \mathcal{I}_{i,n}) \\
&\quad\subseteq \Big(\mathrm{nodes}(F, X^1_{j-1}) \cup \mathcal{K}(F,j-1)\Big) \cap \mathrm{nodes}(F, X^1_i \cap \mathcal{I}_{i,n}) \\
&\quad\overset{(j<i)}{\subseteq} \mathrm{nodes}(F, X^1_{j-1} \cap \mathcal{I}_{i,n}) \cup \Big(\mathcal{K}(F,j-1) \cap \mathrm{nodes}(F, X^1_i \cap \mathcal{I}_{i,n})\Big) \\
&\quad\overset{(1\le j)}{\subseteq} \mathrm{nodes}(F, X^1_{j-1} \cap \mathcal{I}_{i,n}) \cup \Big(\mathcal{K}(F,j-1) \cap \mathrm{nodes}(F, X^j_i \cap \mathcal{I}_{i,n})\Big),
\end{aligned}$$

proving the required result. □

THEOREM 4.5 *Let $F = (\pi, \ldots, O)$ be a BDD over $X_n$, for which we assume the natural variable ordering $\pi$ with $\pi(i) = x_i$ $(1 \leq i \leq n)$. Let $|F_j{}'|$ denote the size of the BDD after moving variable $x_i$ to position $j$. When moving up a variable $x_i \in X_n$, as a lower bound on the size of the resulting BDD $F'$ we have*

$$\begin{aligned}
\mathrm{lb}^{\uparrow}(F, x_i) &= \min_{j=1,\ldots,i-1} |F_j{}'| & (4.1)\\
&\geq \min_{j=1,\ldots,i-1} \{ \mathrm{label}(F, X_{j-1}^1) + \mathrm{label}(F, X_{i-1}^j \setminus \mathcal{I}_{i,n}) \\
&\qquad + \left| \mathcal{K}(F, j-1) \cap \mathrm{nodes}(F, X_i^j \cap \mathcal{I}_{i,n}) \right| \\
&\qquad + \mathrm{label}(F, X_n^{i+1}) \} & (4.2)\\
&= \mathrm{label}(F, X_{i-1}^1 \setminus \mathcal{I}_{i,n}) + \left| O \cap \mathrm{nodes}(F, X_i^1 \cap \mathcal{I}_{i,n}) \right| \\
&\quad + \mathrm{label}(F, X_n^{i+1}). & (4.3)
\end{aligned}$$

Equation (4.1) describing the lower bound $\mathrm{lb}^{\uparrow}(F, x_i)$ states the *exact* lower bound as minimum of the BDD sizes resulting from the movements of variable $x_i$. Of course these exact BDD sizes are not known unless the variable in fact is moved. The goal however is to avoid as many variable swaps as possible. Hence, in Inequality (4.2) and finally in Equation (4.3), the exact lower bound is weakened to a lower bound which prepares a calculation *without* any variable movement (achievable later by a further weakening, see Lemma 4.9).

Inequality (4.2) states a lower bound counting the nodes in the "kernel" $\mathcal{K}(F, j-1)$ as introduced in Section 3.2, which are representing cofactors. To adapt this lower bound to the new context, it has been restricted to count nodes in the middle part, consisting of nodes in levels $j, \ldots, i$. In this, the tightest lower bound known from exact BDD minimization is adapted to the context of dynamic reordering.

Equation (4.3) states a lower bound which only uses output nodes. In this, the bound might appear less tight than that stated in Inequality (4.2). The following proof shows that this is not the case, by proving Equation (4.3). Instead of showing soundness of the final lower bound in line three directly,

a) the soundness of the lower bound expressed in Inequality (4.2), and after this,

b) the equality of the lower bound expressed in Inequality (4.2) and in Equation (4.3)

is shown. Hence, this is a stronger result which implies that no further increase of tightness can be achieved by considering nodes representing cofactors instead of output nodes.

**Proof of Theorem 4.5.** Conceptually, the BDD can be split into three parts: an upper part, consisting of the nodes in levels $1, \ldots, j-1$, a middle part, consisting of the nodes in levels $j, \ldots, i$, and a lower part, consisting of the nodes in levels $i+1, \ldots, n$.

For Inequality (4.2), variable $x_i$ is moved through the middle part only, hence upper and lower part do not change during variable swaps. This is due to a swap being an operation with only local effect in the levels affected. Consequently, the sum of the node numbers in the upper, non-interacting middle and lower part build a first trivial lower bound on the size of the BDD resulting from the variable movement of $x_i$ to position $j$.

In Inequality (4.2), the number of nodes in the following set is added to this (trivial) lower bound: this is the set of nodes residing in middle and lower part which either a) have direct references from nodes in the upper part or b) are output nodes in the middle or lower part, i.e. $\mathcal{K}(F, j-1)$. Before adding the cardinality of this set to the trivial bound, this set is intersected with the set of those nodes in the middle part which interact with $x_i$ (expressed by nodes$(F, X_i^j \cap \mathcal{I}_{i,n})$). Since the nodes in $\mathcal{K}(F, j-1)$ represent the cofactors in cof$(f, X_{j-1}^1)$ by Lemma 3.10, the nodes in this set represent cofactors which

- cannot vanish during movement of $x_i$ (since otherwise the function represented by $F$ would not be preserved) and
- are not already counted otherwise in
  - the part of the trivial lower bound related to the nodes of the middle part which are labeled with a non-interacting variable (since only nodes labeled with an interacting variable are counted) and in
  - the parts of the trivial lower bound related to the upper and lower part of the BDD (since only nodes in the middle part are counted).

Hence, adding these nodes to the trivial lower bound yields a sound and tighter lower bound on the BDD size resulting from the movement of $x_i$.

Equation (4.3) specializes the expression which is minimized in Inequality (4.2) to the case of $j = 1$: this is easily seen because $O$ is the set of output nodes and equals $\mathcal{K}(F, 0)$. Hence, Equation (4.3) claims to obtain the minimum of the expression given in Inequality (4.2) over all $j$ from $1, \ldots, i-1$ in the case of $j = 1$.

First note that, by setting $j$ to 1, i.e. by moving $x_i$ as far as possible, the upper part vanishes. Yet, as the upper part vanishes, the middle part is increased, but it is increased only by levels that belonged to the upper part and whose nodes are labeled with a *non-interacting* variable.

Hence, by moving $x_i$ from position $j$ to 1 in the ordering, the expression in Inequality (4.2) is decreased by the number of interacting nodes in the upper part, i.e. by the term

$$\left|\text{nodes}(F, X_{j-1}^1 \cap \mathcal{I}_{i,n})\right|. \tag{4.4}$$

Moreover, instead of the term $\left|\mathcal{K}(F, j-1) \cap \text{nodes}(F, X_i^j \cap \mathcal{I}_{i,n})\right|$ occurring in Inequality (4.2) we have the term $|O \cap \text{nodes}(F, X_i^1 \cap \mathcal{I}_{i,n})|$ in Equation (4.3).

Together with (4.4), we have an *increase* of

$$\left|O \cap \text{nodes}(F, X_i^1 \cap \mathcal{I}_{i,n})\right|$$

opposing a *decrease* of

$$\left|\text{nodes}(F, X_{j-1}^1 \cap \mathcal{I}_{i,n}) \cup \left(\mathcal{K}(F, j-1) \cap \text{nodes}(F, X_i^j \cap \mathcal{I}_{i,n})\right)\right|.$$

By Corollary 4.4, for every $j$ with $1 \leq j < i$ the stated increase is smaller or equal than the stated decrease.

This yields Equation (4.3) since now, for every $1 \leq j < i$, the expression to minimize over all $j$ from $1, \ldots, i-1$ in Inequality (4.2) must be larger than or equal to the expression of Equation (4.3). □

EXAMPLE 4.6 *In Figure 4.2, let both diagrams be BDDs over $X_n$. An upward movement of variable $x_2$ in the left BDD $F_1$ results in the right BDD $F_2$. Theorem 4.5 predicts a lower bound* $\text{lb}^{\uparrow}(F, x_2)$ *on the size of $F_2$, thereby looking only at $F_1$. Let the sub-graphs $A, B$ and $C$ be non-isomorphic and let* $R := \text{label}(F_1, X_n^3)$. *Hence,* $\text{lb}^{\uparrow}(F_1, x_2) = |X_1^1 \cap \mathcal{I}_{2,n}| + |O \cap \text{nodes}(F_1, X_2^1 \cap \mathcal{I}_{2,n})| + \text{label}(F_1, X_n^3) = 0 + 3 + R = 3 + R$.

*This lower bound exactly predicts the correct BDD size, as the right BDD $F_2$ in fact has $3 + R$ nodes. The minimum number of nodes as given by Theorem 4.2 however is only* $\text{label}(F_1, X_1^1 \setminus \mathcal{I}_{2,n}) + |X_1^1 \cap \mathcal{I}_{2,n}| + \frac{1}{2^{|X_1^1 \cap \mathcal{I}_{2,n}|}} \cdot \text{label}(F_1, x_2) + \text{label}(F_1, X_n^3) = 0 + 1 + \frac{1}{2} \cdot 2 + R = 2 + R$, *thus missing the true number.*

However, there are also examples where the lower bound of Theorem 4.2 predicts larger node numbers than the new lower bound given in

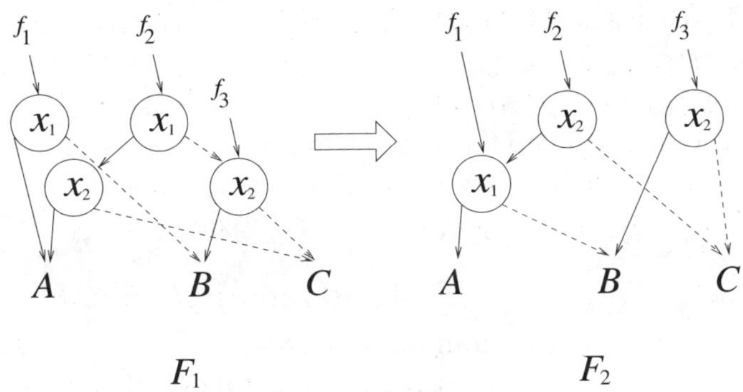


*Figure 4.2.* BDDs for Example 4.6.

Theorem 4.5: this is the case if only few output nodes are opposing many nodes labeled $x_i$ (where $x_i$ is the variable moved upwards). For this reason, in the final suggestion of a new lower bound, the classical and the newly found lower bounds will always be *combined* by building the maximum of the stated node numbers (see later in Section 4.4).

In the following result, another idea from exact BDD minimization is transferred to the context of dynamic reordering: this time the bottom-up lower bound of JANUS described in Section 3.1 is used. This is the tightest lower bound for the bottom-up B&B framework known so far, improving the previous lower bound from [ISY91].

THEOREM 4.7 *Let $F = (\pi, \ldots, O)$ be a BDD over $X_n$ for which we assume the natural variable ordering $\pi$ with $\pi(i) = x_i$ $(1 \leq i \leq n)$. When moving up a variable $x_i \in X_n$, as a lower bound on the size of the resulting BDD $F'$ we have*

$$\text{lb}^{\uparrow}(F, x_i) \quad = \quad |\text{ref}(F, i)| - \left|O_i^1\right| + \text{label}(F, X_n^{i+1}).$$

**Proof.** Let $F' = (\ldots, \ldots, P)$ denote the BDD resulting from the movement of variable $x_i$ and let $F$, $F'$ represent a Boolean function $f$. We have

$$|\text{ref}(F, i)| = |\text{ref}(F', i)| = \left|\text{tcof}(f, X_i^1)\right| \tag{4.5}$$

by Lemma 3.10. By upward movements of variable $x_i$ the lower part of $F$ remains unchanged due to locality of the swap operation $(*)$.

Hence, we have

$$\left|O_i^1\right| = \left|P_i^1\right| \tag{4.6}$$

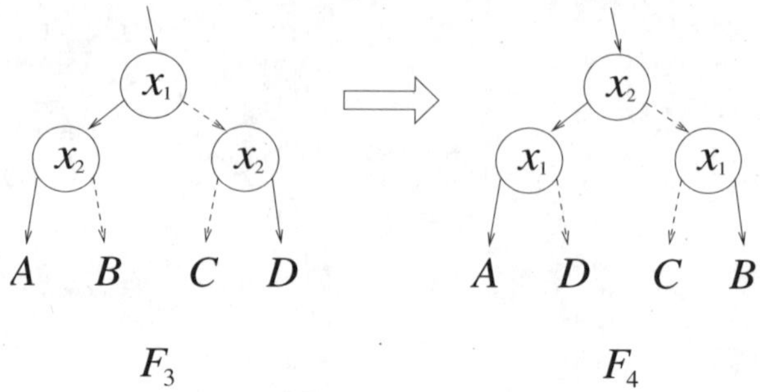


*Figure 4.3.* BDDs for Example 4.8.

since otherwise output nodes would have vanished in the upper part, contradicting that $F$ and $F'$ represent the same function. Applying Lemma 3.5 to $F'$ yields

$$\text{label}(F', X_i^1) \geq |\text{ref}(F', i)| - \left|P_i^1\right|. \tag{4.7}$$

Applying Equations (4.5) and (4.6) to Inequality (4.7) we obtain

$$\text{label}(F', X_i^1) \geq |\text{ref}(F, i)| - \left|O_i^1\right|.$$

By $(*)$ we now have

$$\text{label}(F', X_i^1) + \text{label}(F', X_n^{i+1}) \geq |\text{ref}(F, i)| - \left|O_i^1\right| + \text{label}(F, X_n^{i+1}),$$

which yields the result since $\text{label}(F', X_i^1) + \text{label}(F', X_n^{i+1}) = |F'|$. □

EXAMPLE 4.8 *In Figure 4.3, again let both diagrams be BDDs over $X_n$. An upward movement of variable $x_2$ in the left BDD $F_3$ results in the right BDD $F_4$. Let the sub-graphs $A, B, C$ and $D$ be non-isomorphic. Then the roots of these graphs represent four distinct nodes in* $\text{ref}(F_3, 2)$. *Further let $R :=$* $\text{label}(F_3, X_n^3)$. *Hence, Theorem 4.7 predicts a lower bound* $\text{lb}^{\uparrow}(F_3, x_2) = |\text{ref}(F_3, 2)| - |O_2^1| + \text{label}(F_3, X_n^3) = 4 - 1 + R = 3 + R$.

*Again, this lower bound predicts the correct BDD size, as the right BDD $F_4$ in fact has $3 + R$ nodes. The minimum number of nodes as given by Theorem 4.2 however is only* $\text{label}(F_3, X_1^1 \setminus \mathcal{I}_{2,n}) + |X_1^1 \cap \mathcal{I}_{2,n}| + \frac{1}{2^{|X_1^1 \cap \mathcal{I}_{2,n}|}} \cdot \text{label}(F_3, x_2) + \text{label}(F_3, X_n^3) = 0 + 1 + \frac{1}{2} \cdot 2 + R = 2 + R$, *again staying below the true number.*

## 4.3 Efficient Forms of Improved Lower Bounds

In [DGS01] it has been pointed out that it is crucial to avoid lower bounds whose computation is too time-consuming. Unfortunately, this is the case for the lower bounds presented in Lemma 4.3 and Theorem 4.7: this is due to the terms $|O \cap \text{nodes}(F, X_i^1 \cap \mathcal{I}_{i,n})|$, $|\text{ref}(F,i)|$ and $|O_i^1|$ occurring in the definition of the bounds.

To compute $|O_i^1|$, the number of output nodes in every level has to be maintained with every variable swap. For this, every node in the levels affected by the swap must be checked for membership in $O$, the set of output nodes. This must be done since lists of the numbers of output nodes on every level have to be updated properly. The most efficient way to do this membership test would be a hashing schema which already is too expensive, slowing down the sifting algorithm too much. Of course, the same holds for the term $|O \cap \text{nodes}(F, X_i^1 \cap \mathcal{I}_{i,n})|$ whose computation is even more complicated.

The most efficient method known to compute a "kernel" $|\text{ref}(F,i)|$ involves a traversal of the whole graph (see Section 3.1.2.1), hence actually using this term in a lower bound results in a great loss of performance.

Therefore, the lower bounds given in the previous section must be *weakened* such that

- the soundness of the lower bounds is preserved, and

- the resulting lower bounds only use terms, for which an efficient method of computation is available.

The next results give weakened forms of the previously stated lower bounds, which can be computed efficiently and thus are appropriate for use in dynamic reordering.

LEMMA 4.9 *Let $F = (\pi, \ldots, O)$ be a BDD over $X_n$ for which we assume the natural variable ordering $\pi$ with $\pi(i) = x_i$ $(1 \le i \le n)$. When moving up a variable $x_i \in X_n$, as a lower bound on the size of the resulting BDD $F'$ we have*

$$\begin{aligned}
\text{lb}^{\uparrow}(F, x_i) &= \text{label}(F, X_{i-1}^1 \setminus \mathcal{I}_{i,n}) + \left|O \cap \text{nodes}(F, X_i^1 \cap \mathcal{I}_{i,n})\right| \\
&\quad + \text{label}(F, X_n^{i+1}) \\
&\ge \text{label}(F, X_{i-1}^1 \setminus \mathcal{I}_{i,n}) + |\text{nodes}(F, \{x_1\} \cap \mathcal{I}_{i,n})| \\
&\quad + \text{label}(F, X_n^{i+1}) \\
&= \text{label}(F, X_{i-1}^1 \setminus \mathcal{I}_{i,n}) + \text{label}(F, \{x_1\} \cap \mathcal{I}_{i,n}) \\
&\quad + \text{label}(F, X_n^{i+1}).
\end{aligned}$$

**Proof.** The first line of the equation defining the lower bound $\mathrm{lb}^{\uparrow}(F, x_i)$ gives the result from Lemma 4.3. It is

$$\begin{aligned} & O \cap \mathrm{nodes}(F, X_i^1 \cap \mathcal{I}_{i,n}) \supseteq O \cap \mathrm{nodes}(F, \{x_1\} \cap \mathcal{I}_{i,n}) \\ = \; & \mathrm{nodes}(F, \{x_1\} \cap \mathcal{I}_{i,n}). \end{aligned} \quad (4.8)$$

Equation (4.8) holds: nodes in the uppermost BDD level do not have predecessors, hence nodes in the $x_1$-level must be root nodes. Since all root nodes are output nodes by definition of SBDDs (see Definition 2.12), the equation follows. □

All terms occurring in the last line of the equation giving the weakened lower bound $\mathrm{lb}^{\uparrow}(F, x_i)$ can be computed efficiently, as modern BDD packages maintain level sizes in dedicated variables and interaction tests can be performed using a pre-computed interaction matrix.

Basically, the lower bound stated in Lemma 4.9 restricts the consideration of output nodes to those being roots. Thus, a decrease in tightness must only be expected if there are many *inner* output nodes. However, in practice often many of the output nodes will be situated as roots at the first level of the BDD.

COROLLARY 4.10 *Let $F = (\pi, \ldots, O)$ be a BDD over $X_n$ for which we assume the natural variable ordering $\pi$ with $\pi(i) = x_i$ $(1 \leq i \leq n)$. When moving up a variable $x_i \in X_n$, as a lower bound on the size of the resulting BDD $F'$ we have*

$$\begin{aligned} \mathrm{lb}^{\uparrow}(F, x_i) \; = \; & \mathrm{label}(F, X_{i-1}^1 \setminus \mathcal{I}_{i,n}) + \left| X_{i-1}^2 \cap \mathcal{I}_{i,n} \right| \\ & + \mathrm{label}(F, \{x_1\} \cap \mathcal{I}_{i,n}) + \mathrm{label}(F, X_n^{i+1}). \end{aligned}$$

**Proof.** It is straightforward to see that the term

$$\left| X_{i-1}^2 \cap \mathcal{I}_{i,n} \right| + \mathrm{label}(F, \{x_1\} \cap \mathcal{I}_{i,n}) \quad (4.9)$$

tightens the result from Lemma 4.9 without loss of soundness: at least one node must remain in each level from 2 to $i-1$ after moving $x_i$ upwards. This holds even if moved as far as possible, i.e. if moved to level 1. To avoid collisions of the summands of term (4.9), we have to take care for the following: first since we already count the nodes of the $x_1$-level with the summand $\mathrm{label}(F, \{x_1\} \cap \mathcal{I}_{i,n})$, only the remaining nodes of levels $2, \ldots, i-1$ can be counted in order not to count a node twice. After moving variable $x_i$ upwards to level 1, the output nodes formerly labeled $x_1$ must be contained in the $x_1$-level and the $x_i$-level, both of which do not contribute to the first summand of (4.9). Hence we assume one node per level for those levels only whose nodes are not already counted otherwise. Second, as the non-interacting part of the upper BDD part is already counted in the term $\text{label}(F, X_{i-1}^1 \setminus \mathcal{I}_{i,n})$, we have to intersect with $\mathcal{I}_{i,n}$ in both summands of term (4.9). This again happens to avoid counting nodes twice and to preserve soundness of the lower bound. □

LEMMA 4.11 *Let $F = (\pi, \ldots, O)$ be a BDD over $X_n$ for which we assume the natural variable ordering $\pi$ with $\pi(i) = x_i$ $(1 \leq i \leq n)$. When moving up a variable $x_i \in X_n$, as a lower bound on the size of the resulting BDD $F'$ we have*

$$\begin{aligned}
&\text{lb}^{\uparrow}(F, x_i) \\
&= |\text{ref}(F,i)| - \left|O_i^1\right| + \text{label}(F, X_n^{i+1}) \\
&\geq \left(\text{label}(F, x_{i+1}) - \left|O_n^{i+1}\right|\right) - \left|O_i^1\right| + \text{label}(F, X_n^{i+1}) \qquad (4.10) \\
&= \text{label}(F, x_{i+1}) - |O| + \text{label}(F, X_n^{i+1}).
\end{aligned}$$

**Proof.** The first line of the equation defining the lower bound $\text{lb}^{\uparrow}(F, x_i)$ gives the result from Theorem 4.7. Inequality (4.10) holds: nodes in level $i+1$ are either referenced directly from a node in levels $1, \ldots, i$ or they are output nodes which are contained in $O_{i+1}^{i+1} \subseteq O_n^{i+1}$. Consequently we have $\text{ref}(F, i) \supseteq \text{level}(F, i+1) \setminus O_n^{i+1}$, yielding Inequality (4.10). □

Again, all terms occurring in the last line of the equation giving the weakened lower bound $\text{lb}^{\uparrow}(F, x_i)$ can be computed efficiently. This also holds for the term $|O| = |\text{ref}(F, 0)|$ since the set $\text{ref}(F, 0)$ can be precomputed using one single graph traversal, see Section 3.1.2.1. This traversal always can be used, regardless of the type of BDD application, e.g. VLSI CAD or symbolic state space search. When focusing on VLSI CAD, a good idea would be to use the (often slightly larger) number of output functions pre-declared in the logic level description of a circuit.

Let us again consider the situation illustrated with the BDDs $F_3$ and $F_4$ in Figure 4.3. The lower bound stated in Lemma 4.11 can still exactly predict the correct BDD sizes, but now this is only the case if we assume that the root nodes of the sub-graphs $A, B, C$ and $D$ are all labeled with $x_3$: only then the term $\text{label}(F_3, 3)$ equals four and thus remains as large as the term $\text{ref}(F_3, 2)$ in Example 4.8. Note that in this case this weakened bound still is tighter than the one stated in Theorem 4.2.

## 4.4 Combination of Improved Lower Bounds with Classical Bounds

In this section, the tightest lower bounds for dynamic reordering known so far, given in Theorem 4.2, are combined with the efficient forms of the improved lower bounds presented in this chapter. By this a new, tighter lower bound is obtained.

To use the classical and the improved lower bounds in parallel, it is crucial to carefully avoid counting nodes twice, as this would destroy soundness of the new combined bound.

THEOREM 4.12 *Let $F = (\pi, \ldots, O)$ be a BDD over $X_n$ for which we assume the natural variable ordering $\pi$ with $\pi(i) = x_i$ $(1 \leq i \leq n)$. When moving up a variable $x_i \in X_n$, as a lower bound on the size of the resulting BDD $F'$ we have*

$$\begin{aligned}
&\mathrm{lb}^{\uparrow}(F, x_i) \\
&= \max\{\mathrm{label}(F, X_{i-1}^1 \setminus \mathcal{I}_{i,n}) + \max\{|X_{i-1}^2 \cap \mathcal{I}_{i,n}| \\
&\qquad + \mathrm{label}(F, \{x_1\} \cap \mathcal{I}_{i,n}), |X_{i-1}^1 \cap \mathcal{I}_{i,n}| \\
&\qquad + \frac{1}{2^{|X_{i-1}^1 \cap \mathcal{I}_{i,n}|}} \cdot \mathrm{label}(F, x_i)\}, \mathrm{label}(F, x_{i+1}) - |O|\} \\
&\quad + \mathrm{label}(F, X_n^{i+1}).
\end{aligned}$$

**Proof.** We conceptually divide the BDD into an upper and a lower part. The upper part is partitioned into a set of nodes with labels interacting with $x_i$ ("interacting upper part") and the set of remaining nodes ("non-interacting upper part"). The lower bound is then the sum of a max-term defining a lower bound on the size of the upper part and the size of the lower part which does not change during movement of $x_i$.

The two max-terms "parallelize" classical lower bounds with the improved lower bounds given in the previous sections, such that counting nodes twice by different lower bounds is avoided. Hence, "collisions" of the different lower bounds do not occur and with that the new lower bound must be sound and it is formally tighter than the classical bounds.

In detail: the terms describing (a lower bound on the size of) the non-interacting upper part and the invariant lower part are the well-known "trivial bounds" as also used in the equation describing $\mathrm{lb}^{\uparrow}(F, x_i)$ in Theorem 4.2. The max-term yields the maximum of two terms defining a lower bound on the size of the upper part. The first term is the sum of the trivial lower bound on the non-interacting upper part and a second max-term describing a lower bound on the interacting upper part. In this, the first term is using the partition of the upper part into non-interacting and interacting nodes. The second term is the improved lower bound from Lemma 4.11, describing a lower bound on the size of the upper part as a whole.

With the second max-term we obtain the maximum of

- the well-known lower bound on the size of the interacting upper part stated in Theorem 4.2 and
- the term $|X_{i-1}^2 \cap \mathcal{I}_{i,n}| + \text{label}(F, \{x_1\} \cap \mathcal{I}_{i,n})$,

i.e. the result from Corollary 4.10. □

To see how large the increase of tightness over the best lower bound known so far actually can be, see the following example.

EXAMPLE 4.13 *Let $n \in \mathbb{N}$ be even and let us reconsider a BDD $F$ for the function $f\colon \mathbf{B}^n \to \mathbf{B}$; $(x_1, x_2, \ldots, x_n) \mapsto x_1 \cdot x_2 + x_3 \cdot x_4 + \ldots + x_{n-1} \cdot x_n$, respecting the variable ordering $\pi(1) = x_1, \pi(2) = x_3, \ldots, \pi(n/2) = x_{n-1}$, $\pi(n/2+1) = x_2, \pi(n/2+2) = x_4, \ldots, \pi(n) = x_n$, as given in Figure 4.4.*

*If variable $x_{n-3}$ is moved upwards, Theorem 4.12 yields the following lower bound on the size of the resulting BDDs:*

$$\begin{aligned} \text{label}(F, x_{n-1}) &- 1 + \text{label}(F, \{x_{n-1}, x_2, x_4, \ldots, x_n\}) \\ &= 2^{(n/2)-1} - 1 + 2^{n/2} - 1 \\ &= 2^{(n/2)-1} + 2^{n/2} - 2, \end{aligned}$$

*due to the term* $\text{label}(F, x_{n-1}) - 1$ *being larger than the other arguments of the outer* max*-term in the lower bound of Theorem 4.12. The lower bound given in Theorem 4.2 however yields a lower bound of only*

$$\begin{aligned} &\frac{\text{label}(F, x_{n-3})}{2^{(n/2)-2}} + |\{x_1, x_3, \ldots, x_{n-5}\}| \\ &+ \text{label}(F, \{x_{n-1}, x_2, x_4, \ldots, x_n\}) \\ &= \frac{2^{(n/2)-2}}{2^{(n/2)-2}} + \frac{n}{2} - 2 + 2^{n/2} - 1 \\ &= 1 + \frac{n}{2} - 2 + 2^{n/2} - 1 \\ &= \frac{n}{2} + 2^{n/2} - 2. \end{aligned}$$

*The new lower bound is larger than the classical lower bound by a number of nodes, which is exponential in $n$, the number of input variables.*

## 4.5 Experimental Results

In this section experimental results are given. The experiments have been carried out on a system with an Athlon processor running at 1.4 GHz and a main memory of 1.5 GByte. The classical sifting algorithm

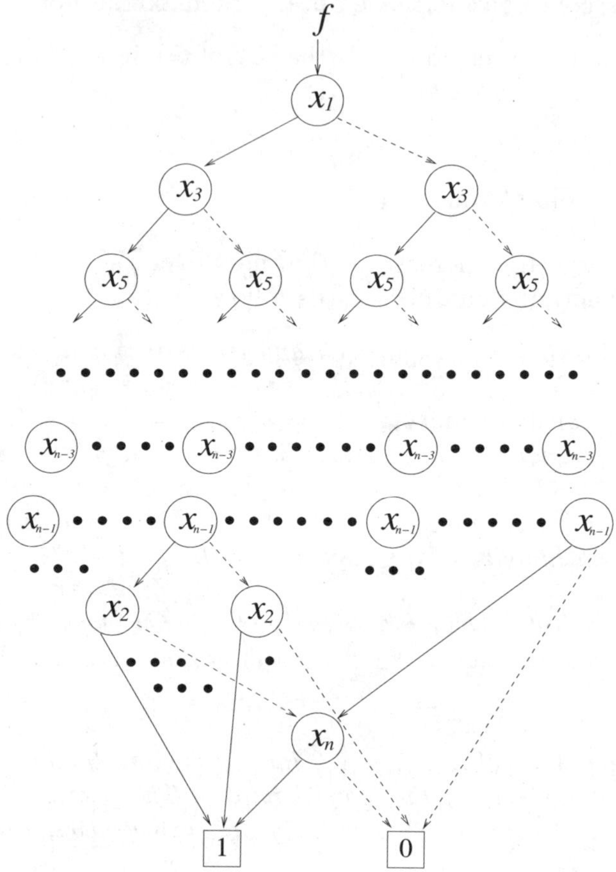


*Figure 4.4.* Example for the tightness of the new lower bound.

without the use of lower bounds is simply called sifting. The classical lower bound sifting approach [DGS01] is called lb-sifting. The recently published enhanced method from [ED05] with the tightened lower bound for moving upwards as presented in this chapter, is called elb-sifting (the lower bound used for moving downwards was that of lb-sifting). The implementation of elb-sifting is based on the sifting routine in the CUDD package [Som02] which already uses a simple way of computing lower bounds, and on lb-sifting. All algorithms have been integrated into the CUDD package, thus running in the same system environment during the experiments.

In two series of experiments all algorithms have been applied to a set of benchmark circuits from LGSynth93 [Col93]. For every benchmark function, a BDD has been built from the logic level description as given

*Table 4.1.* Circuits and initial orderings

| name | in | initial | final |
|---|---|---|---|
| c1355 | 41 | 43869 | 30326 |
| c1908 | 33 | 23158 | 7582 |
| c2670 | 233 | 254562 | 14214 |
| c499 | 41 | 39377 | 30459 |
| c5315 | 178 | 5247 | 2611 |
| c7552 | 207 | 48887 | 12684 |
| c880 | 60 | 15544 | 4548 |
| dalu | 75 | 12946 | 1216 |
| des | 256 | 11835 | 2994 |
| i10 | 257 | 287658 | 98722 |
| i2 | 201 | 334 | 205 |
| i4 | 192 | 420 | 300 |
| i8 | 133 | 3980 | 1678 |
| pair | 173 | 13845 | 5857 |
| rot | 135 | 8322 | 6309 |
| s13207.1 | 700 | 8022 | 3164 |
| s1423 | 91 | 3928 | 1773 |
| s15850.1 | 611 | 40647 | 16591 |
| s38584.1 | 1464 | 100502 | 17371 |
| s5378 | 199 | 5218 | 2405 |
| s9234.1 | 247 | 26984 | 3919 |

in the BLIF file. Table 4.1 gives the name of the function in the first column. Column *in* gives the number of inputs of a function. Column *initial* shows the size of the initial BDD for the function. In Column *final*, the size of the BDD resulting from applying the sifting algorithm is given. Obviously, the resulting BDD sizes are the same for all three sifting approaches: the lower bounds used in the approaches lb-sifting and elb-sifting are sound, i.e. only those parts of the search space are pruned, in which no further improvements are possible.

The results of the first series of experiments are given in Table 4.2. In double column *efficient*, the number of variable swaps (in columns *swaps*) and the run time (in columns *time*) required when using the lower bound of Theorem 4.12 are given. Double column *expensive* states the results for the expensive lower bounds as given in Theorem 4.5 and Theorem 4.7. In column *swaps* the number of variable swaps is given

*Table 4.2.* Comparison of the efficient and the expensive lower bound

| | efficient (Thm. 4.12) | | expensive (Thms. 4.5, 4.7) | | gain | |
|---|---|---|---|---|---|---|
| name | swaps | time | swaps | time | swaps | time |
| c1355 | 1550 | 3.65s | 1538 | 18.37s | -0.8 % | 386.0 % |
| c1908 | 1047 | 1.04s | 1031 | 5.87s | -1.5 % | 414.9 % |
| c2670 | 22634 | 34.93s | 21776 | 530.43s | -3.8 % | 1364.1 % |
| c499 | 1525 | 2.64s | 1523 | 13.81s | -0.1 % | 395.0 % |
| c5315 | 26128 | 0.16s | 24742 | 38.16s | -5.3 % | 23750.0 % |
| c7552 | 44947 | 5.54s | 43929 | 296.13s | -2.3 % | 5284.2 % |
| c880 | 3047 | 0.18s | 2773 | 4.22s | -9.0 % | 2244.4 % |
| dalu | 8075 | 0.24s | 7493 | 12.11s | -7.2 % | 5165.2 % |
| des | 76412 | 0.21s | 75903 | 175.89s | -0.7 % | 83657.1 % |
| i10 | 62701 | 90.12s | 59359 | 2540.63s | -5.3 % | 2723.2 % |
| i2 | 19487 | 0.12s | 19486 | 23.11s | -0.7 % | 17676.9 % |
| i4 | 20936 | 0.05s | 20936 | 27.46s | 0.0 % | 54820.0 % |
| i8 | 23598 | 0.10s | 22517 | 29.37s | -4.6 % | 29270.0 % |
| pair | 33280 | 0.32s | 32330 | 76.17s | -2.9 % | 23703.1 % |
| rot | 15127 | 0.21s | 14612 | 19.99s | -3.4 % | 9419.0 % |
| s13207.1 | 374379 | 1.08s | 354021 | 1500.12s | -5.4 % | 135045.9 % |
| s1423 | 10398 | 0.11s | 9822 | 8.63s | -5.5 % | 7091.7 % |
| s15850.1 | 350888 | 4.25s | 338668 | 2489.01s | -3.5 % | 57649.7 % |
| s38584.1 | 2741100 | 11.92s | 2540187 | 32525.56s | -7.3 % | 275540.3 % |
| s5378 | 37403 | 0.16s | 36441 | 58.29s | -2.6 % | 38760.0 % |
| s9234.1 | 49383 | 1.04s | 47896 | 147.60s | -3.0 % | 13566.7 % |
| $\sum$ | 3924045 | 158.07s | 3676983 | 40540.93s | -6.3 % | 25547.5 % |

and again column *time* states the run time needed. When incorporating the expensive lower bound into the sifting algorithm, these lower bounds have been combined with the existing lower bounds of Theorem 4.2 in a way similar to the combination of lower bounds in Theorem 4.12.

As the results show, only the efficient forms of the lower bounds yield good run times whereas the run times for the expensive forms are far away from being acceptable in practice. Interestingly, the tighter expensive forms of the lower bounds yield a reduction in the number of the swaps which is only 6.3% on average. In this it seems that we do not lose too much pruning power by weakening our bounds to more efficient forms.

*Table 4.3.* Comparison of sifting, lb-sifting and elb-sifting

| | sifting | | lb-sifting (Thm. 4.2) | | elb-sifting (Thm. 4.12) | | gain | |
|---|---|---|---|---|---|---|---|---|
| name | swaps | time | swaps | time | swaps | time | swaps | time |
| c1355 | 3118 | 4.93s | 1620 | 3.81 | 1550 | 3.65s | -4.3 % | -4.2 % |
| c1908 | 2011 | 1.41s | 1101 | 1.10 | 1047 | 1.04s | -4.9 % | -5.5 % |
| c2670 | 69480 | 322.62s | 22983 | 35.59 | 22634 | 34.93s | -1.5 % | -1.9 % |
| c499 | 3123 | 3.75s | 1597 | 2.76 | 1525 | 2.64s | -4.5 % | -4.3 % |
| c5315 | 51583 | 0.41s | 26208 | 0.16 | 26128 | 0.16s | -0.3 % | 0.0 % |
| c7552 | 79264 | 9.27s | 45249 | 5.65 | 44947 | 5.54s | -0.7 % | -1.9 % |
| c880 | 6454 | 1.16s | 3207 | 0.19 | 3047 | 0.18s | -5.0 % | -5.3 % |
| dalu | 10549 | 0.29s | 8211 | 0.23 | 8075 | 0.24s | -1.7 % | 4.3 % |
| des | 103563 | 0.24s | 76566 | 0.21 | 76412 | 0.21s | -0.2 % | 0.0 % |
| i10 | 121245 | 239.79s | 62875 | 97.81 | 62701 | 90.12s | -0.3 % | -7.9 % |
| i2 | 66507 | 0.38s | 19615 | 0.12 | 19487 | 0.12s | -0.7 % | 0.0 % |
| i4 | 64950 | 0.11s | 20985 | 0.06 | 20936 | 0.05s | -0.2 % | -16.7 % |
| i8 | 29711 | 0.12s | 24678 | 0.10 | 23598 | 0.10s | -4.4 % | 0.0 % |
| pair | 56097 | 0.40s | 33545 | 0.32 | 33280 | 0.32s | -0.8 % | 0.0 % |
| rot | 33841 | 0.33s | 15204 | 0.21 | 15127 | 0.21s | -0.5 % | 0.0 % |
| s13207.1 | 804157 | 1.15s | 376807 | 1.07 | 374379 | 1.08s | -0.6 % | 0.9 % |
| s1423 | 15469 | 0.16s | 10412 | 0.12 | 10398 | 0.11s | -0.1 % | -8.3 % |
| s15850.1 | 640002 | 6.49s | 352026 | 4.24 | 350888 | 4.25s | -0.3 % | 0.2 % |
| s38584.1 | 3741639 | 14.88s | 2897064 | 12.58 | 2741100 | 11.92s | -5.4 % | -5.2 % |
| s5378 | 72312 | 0.18s | 37476 | 0.15 | 37403 | 0.16s | -0.2 % | 6.7 % |
| s9234.1 | 112561 | 1.40s | 49415 | 1.05 | 49383 | 1.04s | -0.1 % | -1.0 % |
| $\sum$ | 6087645 | 609.47s | 4086844 | 167.53 | 3924045 | 158.07s | -4.0 % | -5.6 % |

The results of the second series of experiments are given in Table 4.3. In the first column of Table 4.3 the name of the function is given. The next three double columns, *sifting*, *lb-sifting* and *elb-sifting*, state the number of variable swaps (in columns *swaps*) and the run time (in columns *time*) needed for the respective approach. In the last two columns, the gain in percent by using elb-sifting instead of lb-sifting is shown, i.e. in column *swaps* the obtained reductions in the number of performed variable swaps are given and in column *time* the reductions in run time are given.

In the last row, the values given in each single column are accumulated. For the two columns *swaps* and *time* in the double column *gain*, the total gain in percents for the whole test-suite, i.e. the average gain, is given.

As the results in Table 4.3 show, the tighter lower bound is effective, i.e. both the number of variable swaps and the run time are reduced. Using elb-sifting instead of lb-sifting saves up to 5.4% of the variable swaps needed during algorithm run (e.g., see *s38584.1*). On average, the improvement is 4.0%. The obtained reductions in run time are in some cases more than 10% (e.g., see *i4, i10*). On average, a reduction in run time of 5.6% has been obtained. Compared to classical, i.e. unbounded sifting, a reduction in run time of up to 89.2% (*c2670*) is observed. On average, the improvement is 74.1%, i.e. more than a factor of three.

While clearly being significant, the achieved improvements are smaller than those achieved by the use of tightened lower bounds in exact BDD minimization. One reason for this is that the sizes of the search space occurring in these two different fields of application, i.e. in exact minimization and dynamic reordering, differ drastically. The search space that has to be maintained during exact BDD minimization is of a size which is exponential in the number of variables (see Section 3.2.4.2), while the search space that has to be explored while moving one particular variable in the sifting algorithm is only linear in the number of variables (yielding a total search space for all variables which is only of quadratic size).

As can be deduced from previous results with approaches to exact BDD minimization (see Chapter 3), the pruning power of the lower bounds tend to be higher if applied to state spaces of very large sizes.

A second, less important reason is the following: in the course of an appropriate adaption from exact BDD minimization to dynamic reordering, it is necessary to weaken the lower bounds. Otherwise the computation of the original lower bounds as given in Lemma 4.3 and Theorem 4.7 would be too time-consuming when used in an algorithm as fast as sifting. The situation is different in an algorithm with high run times like exact BDD minimization: here, other operations carried out during the algorithm run are of much higher complexity than the computation of the lower bounds (e.g., BDD reconstruction in exact BDD minimization). Consequently, the reduction in run time due to the higher pruning power of tighter bounds exceeds the additional computational effort caused by the higher complexity of the lower bounds.

However, as can be seen from the experiments shown in Table 4.3 and in Table 4.2, the weakened lower bounds presented are still effective in the much smaller search space of dynamic reordering. In this, a widespread and very popular algorithm for heuristic BDD minimization, the sifting algorithm, can still be noticeably accelerated.

## 4.6 Summary

In this chapter classical and recently published lower bounds on BDD sizes for use in dynamic reordering have been presented. The latest lower bounds in part have been adapted from exact BDD minimization as described in Chapter 3, Section 3.2. The theory behind these new lower bounds has been presented and examples have been given, which gave a deeper insight in the effect of lower bounds on the efficiency of dynamic BDD minimization. After presenting the improved lower bounds, proofs of soundness and increased tightness over the classical lower bounds have been given. Next, following the constraints given in practice, the improved lower bounds have been restricted to more efficient forms. The use of the resulting lower bounds during dynamic BDD minimization has been studied experimentally.

Experimental results with benchmark functions show that reductions in run time of almost 90% have been obtained when comparing to classical, unbounded sifting. In comparison to lb-sifting, i.e. sifting using the classical lower bound, the improved lower bound still yields further reductions in run time of more than 10%. This is achieved while preserving full quality of the results.

# Chapter 5

# PATH MINIMIZATION

The classical criterion for the optimality of a BDD is its *size*, i.e. the number of nodes in the diagram. This criterion has been addressed in the last two chapters both with *exact* and *heuristic* approaches. In this chapter, *alternative* criteria for BDD optimality are considered. These criteria are related to *paths* in BDDs.

The number of paths in BDDs as well as their expected and average length have been formally defined in Section 2.4.8.

When moving from the minimization of BDD size to a minimization with respect to path criteria, a first obvious question is: in contrast to the previous minimization of the number of nodes, how can a minimization with respect to the number of paths be done in BDDs? And: are these two criteria independent (orthogonal) optimization objectives, or are they related (correlated) to one another?

The optimization with respect to the number of paths is motivated by a number of applications in VLSI CAD in different areas and it has been studied in [FD02a].

**BDD Circuits.** As has been described in Chapter 1, BDDs can directly be mapped into circuits by replacing BDD nodes with multiplexors. These circuits are well testable [DSF04]. In particular test pattern generation with respect to the path delay fault model is efficient for BDD circuits. The number of paths in the circuit corresponds to the number of paths in the BDD and is proportional to the number of faults under the path delay fault model. Therefore minimizing the number of paths can significantly reduce the time for testing.

**DSOP Minimization.** A *Disjoint Sum of Products* (DSOP) is a representation of a function as a sum of pairwise disjoint products. DSOPs are used in several applications in the area of CAD, e.g. the calculation of spectra of Boolean functions [Fal93, FC99, TDM01] or as a starting point for the minimization of *Exclusive-Or-Sum-Of-Products* (ESOPs), e.g. in [Sas93, MP01]. Some techniques for minimization of DSOPs working on explicit representations of cubes have been introduced in [FSC93, ST02], but are only applicable to small instances of the problem.

From a BDD of a Boolean function $f$ a DSOP can easily be derived: Each path to **1** corresponds to a (partial) assignment to the variables of $f$. This directly corresponds to a product over the literals of $f$. The products retrieved from different paths to **1** are disjoint. Therefore the sum of the products collected from all paths to **1** is a DSOP of $f$. In [FD02b] this technique has been applied. Experimental results have shown that the size of the retrieved DSOPs is equal to that of the other approaches. But due to the implicit representation of the products much larger functions can be handled.

**SAT.** Solving the *satisfiability problem* (SAT problem) for a given Boolean function $f(x_1, \ldots, x_n)$ means to decide whether there exists an assignment such that $f$ evaluates to one, or to proof that no such assignment exists. A large number of problems is solved by reducing the original problem to a SAT problem. Sophisticated algorithms to solve SAT problems are known [MSS96, MMZ$^+$01]. They usually work on a representation of the function in *Conjunctive Normal Form* (CNF).

The relation of the number of paths in BDDs to SAT-solving is twofold:

- The correspondence to the number of back-tracks in a SAT-solver has been shown in [RDO02]. This important interrelation is studied in detail in Chapter 6.

- The number of paths to **0** is equal to the number of clauses in a CNF, if a BDD is mapped to a CNF directly, i.e. without logic optimizations. For example in [CNQ03, GGW$^+$03] techniques to integrate SAT and BDDs were introduced. There BDDs were used as a starting point for the generation of a CNF.

As a first criteria alternative to diagram size, minimizing the number of paths in BDDs is studied in this chapter. Theoretical studies show the exponential dependency on the variable ordering and several instructive examples are given. The results given here also provide answers to the above questions that arose from the shift to the new criterion for BDD optimality. A heuristic technique to minimize the number of paths in

BDDs is introduced. This technique is applied to sifting (see Section 2.4.7.2) to carry out the minimization. A detailed explanation is given and the efficiency is underlined by experiments. It is demonstrated that the number of paths can be significantly reduced for some benchmarks. At the same time the number of nodes does not necessarily increase and may even be also reduced. While the experiments show the feasibility of the approach, it also turns out that minimizing the number of paths can be much more time-consuming than minimizing the number of nodes.

On the other hand, this increased algorithmic hardness is not present for a second path-related criterion, the *Expected Path Length* (EPL). An approach to minimize the expected path length in BDDs has first been suggested in [LWHL01]. However, this approach still had a complexity comparable to the approach for minimizing the number of paths.

Tackling this problem of high run time, two sifting modifications minimizing the expected path length have been published independently [NMSB03, EGD04b] which have the same asymptotic run time complexity as the original sifting algorithm. Minimization of the expected path length is motivated by applications in two areas of VLSI CAD.

**Functional Simulation.** In many verification tools methods for functional simulation, BDDs or free BDDs are used [CG85, AM95, MMS+95, SDB97, LWA98, LWHL01, ISM03, NMSB03, JMB03]. A crucial point here is the time needed to evaluate a function using a representing BDD. The evaluation time is proportional to the expected path length in the BDD (see Section 2.4.8). Thus, in BDD-based functional simulation, the BDDs yielded by the proposed methods to minimize the EPL can reduce the run time of the simulation significantly, because logic functions are repeatedly evaluated with different input vectors.

Boolean functions $f$: $\mathbf{B}^n \to \mathbf{B}^m$ can be viewed as a Boolean relation representable by its *Characteristic Function* (CF). Using a BDD for the CF of this relation, the evaluation time is $O(m+n)$, see [AM95] and [SDB97]. The time complexity of function evaluation using shared BDDs directly representing the function $f$ is higher, $O(m \cdot n)$ [AM95, SDB97]. However, in functional simulation based on BDDs, shared BDDs directly representing the function $f$ often have to be used instead of BDDs for CFs since the sizes of BDDs for CFs tend to be too large to preserve practicality [ISM03]. For this reason, there is a strong demand for algorithms to minimize the EPL, speeding up the evaluation time for shared BDDs.

**Logic Synthesis.** Classically, BDD optimization was done with respect to the number of nodes, i.e. BDD size. This reduces the memory and run time needed for the representation and manipulation of Boolean functions. If BDDs are directly mapped into multiplexor circuits (see Chapter 1), minimization of BDD size directly transfers to a smaller chip area, but it can lead to chips with poor timing performance, see [LAB98]. Recently, approaches based on Rudell's sifting algorithm have been proposed which optimize BDDs targeting the delay of the resulting circuits, see [LAB98] and [SB00]. But these techniques either fail to achieve strictly local operations during variable swaps [LAB98], resulting in high run times, or they apply simplified cost functions [SB00] and thus can produce results which are far away from the true optimization objective.

Minimization of the EPL is also suggested for path delay minimization in multiplexor circuits: assuming a *unit delay model* [DG02], the maximal path length (MPL) in BDDs (see Example 2.27) models the delay on a critical path. In Section 5.2.5 experiments are described showing that the MPL of the BDDs yielded by minimization of the EPL is almost the same as for BDDs directly minimized towards a smallest MPL. When minimizing the EPL, experiments show that reductions in the length of critical paths of more than 40% can be observed on benchmark functions. While the same results are obtained with an approach oriented towards the maximum path length the EPL-oriented method is faster by up to two orders of magnitude.

It also has been proposed to improve the quality of functional decomposition (see Section 2.3 and, in particular, [YC02]) by minimizing the expected path length in BDDs representing the logic functions [NMSB03].

In this chapter new techniques for BDD minimization with respect to the *expected path length* are presented. These techniques are fast heuristic approaches based on sifting and, unlike previous approaches, perform variable swaps with the same time complexity as the original sifting algorithm. Experimental results show speed-ups of up to two orders of magnitude while preserving high quality of the results.

Comparing the run times of the sifting approaches for minimizing the number of paths and of those targeting at the EPL, a significant difference can be seen: in BDDs, it turns out to be much harder to minimize the number of paths than to minimize the EPL.

An obvious question is that for the reasons behind this gap. Next in this chapter, a *unifying view* is applied to compare minimization of the EPL to the approaches to minimize

a) the number of paths in BDDs, and

b) the sum of the lengths of the paths in BDDs.

This allows for a characterization of the different problems in terms of algorithmic hardness. Finally, an algorithm for the minimization of the *Average Path Length* (APL) in BDDs can be derived directly from the approaches targeting the criteria a) and b).

As a side-effect of minimizing the APL in BDDs, the length of the longest path in the circuit is also reduced. Therefore the minimization of the APL leads to circuits with a smaller delay on the critical path. By this, the circuit is optimized for speed [FSD04]. Minimizing the EPL reduces the average evaluation time for the evaluation of a BDD. This takes into account, how often a path is chosen for an input vector applied to the BDD by the environment or to the corresponding circuit, respectively. In this, the APL is defined independently of the environment of the circuit and seems to have a more direct relation to the circuit structure and depth. Experimental results are given showing the feasibility of the approach and the effectiveness of an optimization.

The chapter is organized as follows: starting with the number of paths in BDDs, first theoretical studies are given in Section 5.1.1. The efficient minimization algorithm is explained in Section 5.1.2.

In Section 5.2, the study is continued with the EPL. First, basic techniques to minimize the EPL are discussed. Section 5.2.2 then describes a recent approach to minimize the EPL in BDDs by a sifting method. A technique to keep track of local changes during variable swaps is introduced as the framework for the new approach. Next, in Section 5.2.3, attention is turned to node weights that are used in the approach, explaining a probabilistic interpretation. It is clarified how to compute the weights and a useful invariance property is deduced. The presentation of the algorithm is completed by a detailed schema for node updates in Section 5.2.4. This schema describes what quantities must be refreshed depending on the situation in the algorithm.

In Section 5.3 an algorithm to minimize the APL in BDDs is given. First, in Section 5.3.1 a unifying view of BDD optimization problems is used to derive an approach to minimize the sum of the path lengths in BDDs. Next, in Section 5.3.2 the algorithm to minimize the APL in BDDs is given with the help of this newly obtained method.

For all approaches experimental results are given.

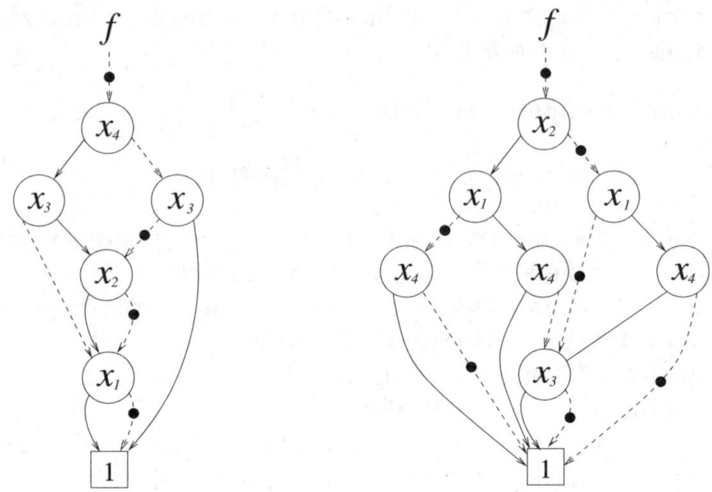


(a) Minimal number of nodes (b) Minimal number of paths to one

*Figure 5.1.* Examples of BDDs.

## 5.1 Minimization of Number of Paths

In this section minimizing the number of paths in BDDs is studied. Theoretical studies are carried out and an algorithm for path minimization is presented.

### 5.1.1 Theory

An introductory examples briefly clarify the difference between representing a function with BDDs of minimal path number or minimal size. Then, the exponential dependence of the number of paths on the variable ordering is shown.

#### 5.1.1.1 Examples

In Example 2.32 it has been demonstrated that all BDDs representing the EXOR-function have a number of nodes linear in $n$ whereas the number of paths is exponential in $n$.

EXAMPLE 5.1 *On the other hand the function* $f = \overline{x}_1\overline{x}_2\overline{x}_3 + \overline{x}_1 x_2 x_4 + x_1 x_2 \overline{x}_3 \overline{x}_4 + x_1 \overline{x}_2 x_3 x_4$ *is an example where different variable orderings lead to a BDD with either a minimal number of nodes (Figure 5.1(a)) or a minimal number of paths to one (Figure 5.1(b)). Note, that the implicit edge representing f is complemented. While the BDD minimal in size has six nodes (including the terminal) and five* **1***-paths, the BDD minimal in the number of* **1***-paths has two more nodes, but only four* **1***-paths. This relation has already been shown in [DM99].*

#### 5.1.1.2 Dependency on the Variable Ordering

As has been demonstrated in Example 2.19, the variable ordering heavily influences the number of nodes of a BDD. The left BDD in Figure 2.8 has $n$ inner nodes, but the right BDD, respecting a different ordering, has an exponential number of nodes. The number of **1**-paths is exponential in $n$ in both cases.

Nonetheless the same influence of the variable ordering can be proven for the number of **1**-paths in a BDD as the following lemma shows.

LEMMA 5.2 *Let the Boolean function $f_n(x_1, \ldots, x_{2n}), n > 1$ be recursively defined:*

- $f_1 = x_1 x_{n+1}$
- $f_j = f_{j-1} + (\bigcap_{i=1}^{j-1} \overline{x}_i) x_j x_{n+j}$

*And let $\pi_1$ and $\pi_2$ be two variable orderings:*

- $\pi_1 = (x_1, x_{n+1}, x_2, x_{n+2}, \ldots, x_n, x_{2n})$
- $\pi_2 = (x_{n+1}, x_1, x_{n+2}, x_2, \ldots, x_{2n}, x_n)$

*The* BDD$(f_n, \pi_1)$ *without CEs with respect to $\pi_1$ has $2n + 2$ nodes and $n$* **1***-paths. But* BDD$(f_n, \pi_2)$ *without CEs with respect to $\pi_2$ has $3n + 1$ nodes and $2^n - 1$* **1***-paths.*

**Proof.** At first the variable ordering $\pi_1$, i.e. the BDD$(f_n, \pi_1)$ is considered (for an easier understanding of the proof see the example for $n = 4$ in Figure 5.2). The nodes in BDD$(f_n, \pi_1)$ correspond to cofactors in $f_n$. By construction any cofactor $f_n|_{x_1=c_1,\ldots x_{j-1}=c_{j-1},x_j=1}$ where $(c_1, \ldots c_{j-1}) \neq (0, \ldots, 0)$ is independent of $x_j, \ldots, x_n$ and $x_{n+j}, \ldots, x_{2n}$. Furthermore

$$f_n|_{x_1=0,\ldots,x_{j-1}=0,x_j=1} = x_{n+j}.$$

Thus, for a node $v$ labeled $x_j$ always var(then($v$)) $= x_{n+j}$ holds, i.e. the sub-graph representing the positive cofactor has only one node labeled $x_{n+j}$. The negative cofactor with respect to $x_j$, $f_n|_{x_1=0,\ldots,x_{j-1}=0,x_j=0}$, is independent of $x_{n+j}$, so var(else($v$)) $= x_{j+1}$. This cofactor can be constructed from $f_n|_{x_1=0,\ldots x_j=0,x_{j+1}=1}$ and $f_n|_{x_1=0,\ldots x_j=0,x_{j+1}=0}$. Now the same argumentation can be applied for cofactorization with respect to $x_{j+1}$. Therefore exactly one node in BDD$(f_n, \pi_1)$ corresponds to each

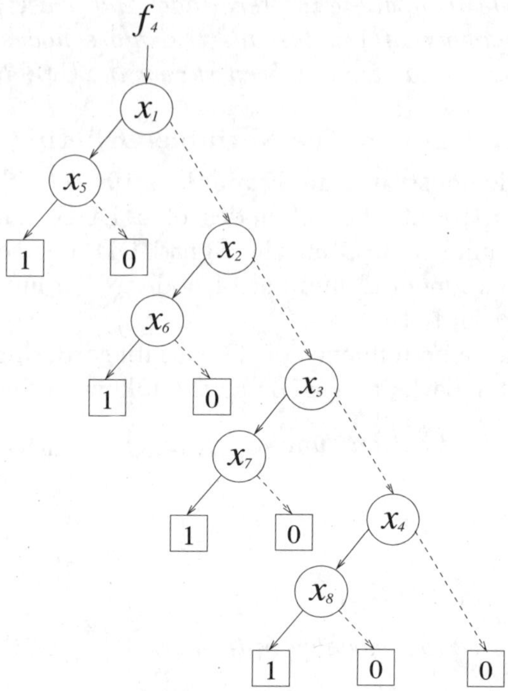


*Figure 5.2.* BDD$(f_4, \pi_1)$, variable ordering $\pi_1$.

variable. Including the two terminal nodes the BDD has $2n + 2$ nodes. Since only cofactors of the form $f_n|_{x_1=0,\ldots,x_{j-1}=0,x_j=1,x_{j+1}=1}$ equal function one, there are $n$ **1**-paths in BDD$(f_n, \pi_1)$.

The transformation from variable ordering $\pi_1$ to $\pi_2$, i.e. BDD$(f_n, \pi_1)$ into BDD$(f_n, \pi_2)$ (see Figure 5.3 for an example) takes $n$ swap operations of pairs $(x_i, x_{i+n})$. Due to the canonicity of BDDs the order of swaps is arbitrary. Let therefore be $(x_n, x_{2n})$, ..., $(x_1, x_{n+1})$ the order of pairs. The swap operations are local operations and are of the type that is shown in Figure 5.4. Before the swap operation there are

$$\mathcal{P}_1(\text{BDD}(g_{i-1}, \pi_1)) = 1 + \mathcal{P}_1(\text{BDD}(g_i, \pi_1))$$

**1**-paths from the topmost node involved, afterwards there are

$$\mathcal{P}_1(\text{BDD}(g_{i-1}, \pi_2)) = 1 + 2 \cdot \mathcal{P}_1(\text{BDD}(g_i, \pi_1))$$

**1**-paths from the topmost node involved in the swap as can be seen in the figure.

Swapping $(x_n, x_{2n})$ does not change the number of paths since $g_n$ is the function zero (i.e. BDD$(g_n, \pi_1)$ = BDD$(g_n, \pi_2) = G_n$), the number

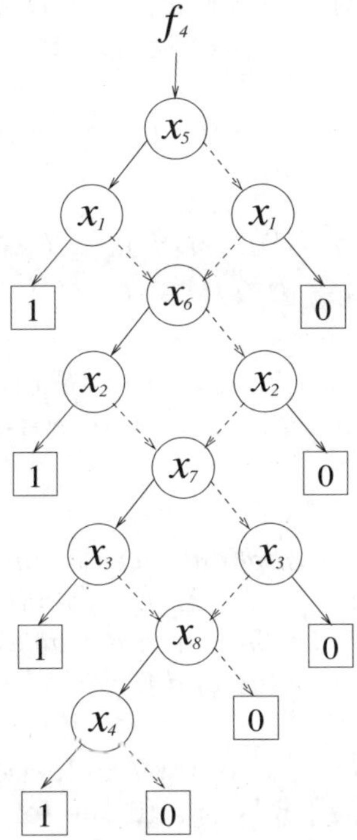


*Figure 5.3.* BDD($f_4, \pi_2$), variable ordering $\pi_2$.

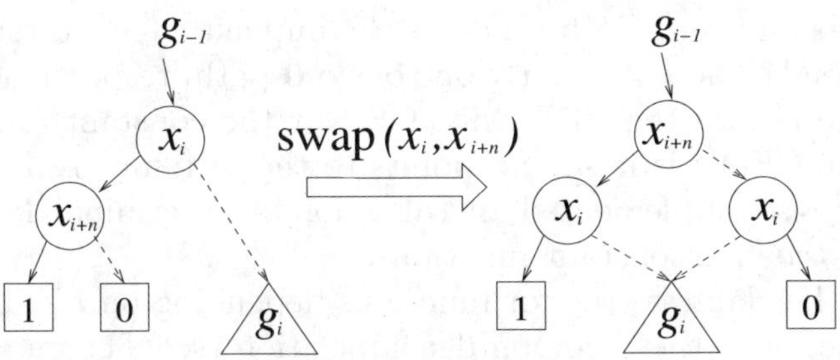


*Figure 5.4.* The swap operation applied to a pair $(x_i, x_{i+n})$ of variables.

of **1**-paths in the sub-graph is $\mathcal{P}_1(\text{BDD}(g_{n-1}, \pi_2)) = 1$. Proceeding with $(x_{n-1}, x_{2n-1})$ results in

$$\mathcal{P}_1(\text{BDD}(g_{n-2}, \pi_2)) = 1 + 2 \cdot \mathcal{P}_1(\text{BDD}(g_{n-1}, \pi_2)).$$

Recursively applying this scheme $\mathcal{P}_1(\text{BDD}(g_0, \pi_2))$ can be calculated:

$$\mathcal{P}_1(\text{BDD}(g_0, \pi_2)) = 1 + 2 \cdot (\ldots 1 + 2 \cdot (1 + \mathcal{P}_1(G_n)) \ldots)$$

Knowing that $g_n$ = zero the result is $\mathcal{P}_1(\text{BDD}(g_0, \pi_2)) = 2^n - 1$. Since $g_0 = f_n$ the BDD of $f_n$ with respect to $\pi_2$ has $2^n - 1$ **1**-paths. □

The same dependence is true for BDDs with CEs. The argument relies on the following two lemmas.

LEMMA 5.3 *Given are two BDDs without CEs $G_f$ and $G_{\overline{f}}$ of functions $f$ and $\overline{f}$, respectively. Then, $P_0(G_f) = P_1(G_{\overline{f}})$ and $P_1(G_f) = P_0(G_{\overline{f}})$.*

**Proof.** The graphs of both BDDs are isomorphic, only the terminal nodes **1** and **0** are exchanged. Therefore a **1**-path in $G_f$ becomes a **0**-path in $G_{\overline{f}}$ and vice versa. □

LEMMA 5.4 *Given are two implicit edges $e$ and $e'$ into BDDs $G_f$ and $G'_f$ of a function $f$ where $G_f$ is a BDD without CEs and $G'_f$ is a BDD with CEs. If both BDDs have the same variable ordering $\pi$, both BDDs have the same number of **1**-paths and **0**-paths.*

**Proof.** The proof is carried out by induction over the number of variables the function $f$ depends on. In the following only the number of **0**-paths is considered, the argument for the number of **1**-paths is analogous.

The BDDs without CEs have only one node for each of the constant functions one and zero. In both cases the implicit edge $e$ directly points to the terminal. There is exactly one or no **0**-path, respectively. For the implicit edge $e'$ into the BDD with CEs for the constant function one, the attribute CE($e'$) is false and points to the terminal **1**. For function zero the edge is complemented and also points to terminal **1**. Therefore the claim is true for constant functions.

Assume, the claim is true for functions depending on $n-1$ variables. Given a function $f$ the Shannon decomposition (see Theorem 2.9) with respect to the topmost variable $x$ in the order returns $f = \overline{x}f_0 + xf_1$ where $f_0$ and $f_1$ are the negative and positive cofactors, respectively. Both functions depend on $n-1$ variables. At most one node is introduced into the BDD at the topmost level. There occur two cases:

1) The new node in $G'_f$ directly represents the function $f$, thus CE($e'$) is false. Then,

$$\begin{aligned} P_0(G_f) &= P_0(G_{f_0}) + P_0(G_{f_1}) \\ &= P_0(G'_{f_0}) + P_0(G'_{f_1}) \\ &= P_0(G'_f) \end{aligned}$$

2) The new node in $G'_f$ represents $\overline{f}$, thus CE($e'$) is true. Then,

$$\begin{aligned} P_0(G_f) &= P_1(G_{\overline{f}}) \qquad \text{(from Lemma 5.3)} \\ &= P_1(G_{\overline{f}_0}) + P_1(G_{\overline{f}_1}) \\ &= P_1(G'_{\overline{f}_0}) + P_1(G'_{\overline{f}_1}) \\ &= P_1(G'_{\overline{f}}) \qquad \text{(due to the CE)} \\ &= P_0(G'_f), \end{aligned}$$

completing the proof. □

In summary, the number of paths in BDDs may drastically change with the variable ordering. Even an exponential change can occur while the number of nodes remains linear in the number of input variables. This shows the potential for an algorithm to minimize the number of paths.

### 5.1.2 Efficient Minimization

This section proposes the technique to minimize the number of paths in BDDs. As efficient BDD packages use CEs, also CEs are taken into account. Basically the technique relies on swapping of adjacent variables and keeps track of changes in the number of paths due to the swap. Then, the integration into Rudell's sifting algorithm is shown.

#### 5.1.2.1 Swapping Variables

Swapping two adjacent variables $x_i$, $x_j$ in the variable ordering $\pi$ results in the new variable ordering $\pi'$ with

$$\pi'(k) = \begin{cases} \pi(k) & \text{if } k \neq i, j \\ \pi(i) & \text{if } k = j \\ \pi(j) & \text{if } k = i \end{cases} .$$

Let in the following $G^\pi := \text{BDD}(f, \pi)$ and $G^{\pi'} := \text{BDD}(f, \pi')$.

The swap operation has only local effects on the BDD regarding the number of nodes and is therefore used in most of the common reordering routines for BDDs. The operation influences the nodes in the two levels $i$ and $j$ but leaves all other nodes untouched. This way it is easy to keep track of the number of nodes in the BDD. One case that occurs during the swap of levels is illustrated in Figure 5.5. Due to the swap of $x_i$ and $x_j$ three nodes instead of the previous two are needed.

While swapping two variables is a local operation when the number of nodes is considered (see Section 2.4.7.1), this is not the case when calculating the number of paths. If the number $p_1(w)$ of regular paths from the outputs to a node $w$ and also the number of regular paths from $w$ to **1** are known, the number of paths from the outputs to **1** via $w$

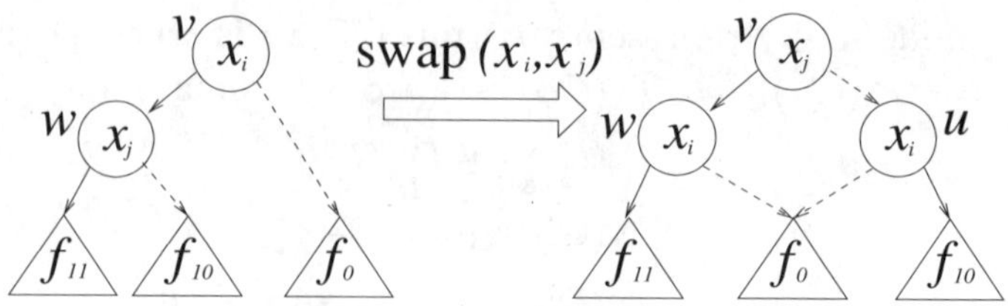


*Figure 5.5.* Nodes in a swap operation.

is simply the product of both. (Of course, to get the total number of **1**-paths via $w$ one has to also add the number of complemented paths to $w$ multiplied by the number of **0**-paths from $w$). By keeping track of all changes, $P_1(G^{\pi'})$ can be calculated, if $P_1(G^{\pi})$ is also given. But a change in $p_1(w)$ will affect $p_1$ of all nodes on a path from $w$ to **1** as well and updating all these values is necessary for the efficient calculation of $P_1$ during further swap operations.

Therefore $P_1(G^{\pi'})$ is calculated by keeping track of all the changes of $p_1(w)$ for all nodes $w$ in levels below $\pi(i)$ during the swap. These changes are propagated down to the terminal **1** afterwards. Obviously the number of **1**-paths from the outputs to **1**, $P_1(G^{\pi'})$, is equal to $p_1(\mathbf{1})$ after the propagation.

For every node $w$ of the BDD the values $p_0(w)$ and $p_1(w)$ can be calculated from values $p_0$ and $p_1$ of all predecessors of $w$:

$$p_0(w) = \sum_{v \in \mathcal{M}_1(w)} p_0(v) + \sum_{v \in \mathcal{M}_0(w)} p_1(v) \tag{5.1}$$

$$p_1(w) = \sum_{v \in \mathcal{M}_1(w)} p_1(v) + \sum_{v \in \mathcal{M}_0(w)} p_0(v) \tag{5.2}$$

At first it is shown how to keep track of the changes and then how the propagation is done.

#### 5.1.2.2 Keeping Track of Changes

For each node $d_1(t)$ ($d_0(t)$) denotes the difference in the number of regular (complemented) paths from the outputs to $t$ before and after the swap. Looking at Figure 5.5 again one can see the changes in the number of paths. The number $p_0(w)$ ($p_1(w)$) does not change and therefore $d_0(w)$ ($d_1(w)$) is not changed either. Before the swap the node $u$ did not have $v$ as a predecessor in previous calculations of $p_0(u)$ ($p_1(u)$), so the following updates are done:

$$p_0(u) := p_0(u) + p_0(v) \quad (p_1(u) := p_1(u) + p_1(v))$$

and

$$d_0(u) := d_0(u) + p_0(v) \quad (d_1(u) := d_1(u) + p_1(v))$$

Figure 5.5 is just one case of the changes made to nodes in levels below the $x_i$-level but the schema of the updates applies to all other cases as well. By iterating the nodes of the $x_i$-level all the changes are accumulated in the values $d_0$ and $d_1$ of nodes in levels below (consider the case resulting from going through the example in Figure 5.5 from right to left).

For efficiency a stack $s(k)$ is assigned to each level $k$ of the BDD. Every time a value $d_0(v)$ or $d_1(v)$ is changed the corresponding node $v$ is pushed onto the stack corresponding to var($v$), i.e. stack $s(k)$ where $\pi'(k) =$ var($v$). Thereby only those nodes are considered during propagation where a change in the number of paths may have occurred. This method to keep track of changes does not change the asymptotic time complexity of the swapping operation since only a constant number of operations are added, these are additions and lookups on nodes that are touched anyway.

#### 5.1.2.3 Propagation of Changes

Looking at Equations (5.1) and (5.2) the differences $d_0(w)$ and $d_1(w)$ can be calculated from the differences of all predecessors of $w$, respectively:

$$d_0(w) = \sum_{v\in\mathcal{M}_1(w)} d_0(v) + \sum_{v\in\mathcal{M}_0(w)} d_1(v) \qquad (5.3)$$

$$d_1(w) = \sum_{v\in\mathcal{M}_1(w)} d_1(v) + \sum_{v\in\mathcal{M}_0(w)} d_0(v) \qquad (5.4)$$

The algorithm works in a different way since efficient implementations of BDD packages only store information about the children of a node, but not of its predecessors. All nodes in the stacks are candidates for a change in the number of paths leading to the node from the outputs. Figure 5.6 shows the algorithm to propagate the changes. For simplicity only the algorithm to handle BDDs without CEs is shown, the extension to handle CEs is straightforward and has been implemented. Subroutine **pop_from_stacks** returns in $v$ the first element in the topmost stack that is not empty. Here, topmost means associated to the highest level. This leads to a level-wise propagation of changes. All predecessors of a node $v$ are visited before the node itself. Thus, Equations (5.3) and (5.4) have been evaluated correctly when $v$ is visited.

It is checked if $d_0(v)$ or $d_1(v)$ is different from zero, because $v$ might be on the stack twice (when pushing $v$ onto the stack it is not checked if $v$ is an element of the stack already). Then, the counts for the children of $v$ are updated by calling **update_counts**. If a child is a terminal node no further propagation is necessary. Otherwise **push_onto_stacks** pushes the child node onto the stack corresponding to the node's level in the BDD. After propagating all updates from a node $v$, $d_0(v)$ and $d_1(v)$ have

```
propagate_changes()
    proc
        while pop_from_stacks(&v) do
            if d0(v) ≠ 0 or d1(v) ≠ 0 then
                w := then(v);
                update_counts(v,w);
                if not IsTerminal(w) then
                    push_onto_stacks(w);
                end–if
                w := else(v)
                update_counts(v,w);
                if not IsTerminal(w) then
                    push_onto_stacks(w);
                end–if
                d0(v) := 0;
                d1(v) := 0;
            end–if
        end–while
    end–proc

update_counts(node v, node w)
    proc
        p0(w) := p0(w) + d0(v);
        d0(w) := d0(w) + d0(v);
        p1(w) := p1(w) + d1(v);
        d1(w) := d1(w) + d1(v);
    end–proc
```

*Figure 5.6.* Algorithm to propagate changes.

to be set to zero before visiting the node a second time and before the next swap operation.

Despite the application of calculating the number of **1**-paths this technique also calculates the number of **0**-paths, $P_0(G^\pi)$ in $p_0(\mathbf{1})$, and the number of all paths in the BDD which is given by $P_0(G^\pi) + P_1(G^\pi)$.

#### 5.1.2.4 Modification of Sifting

With a swapping procedure that calculates the number **1**-paths Rudell's sifting algorithm can be modified to minimize the number of **1**-paths instead of the size of a given BDD. In the original algorithm every variable is moved up and down in the variable ordering. At each position the size of the BDDs is measured and finally the variable is fixed at a position where the BDD was smallest. All the changes in the variable ordering are done by swapping adjacent variables.

If the described swapping procedure is used, the number of **1**-paths in the BDD can be used as the criterion which position to choose for a variable rather than the size of the BDD. After each swap of two adjacent variables, i.e. after processing all nodes in the two levels, changes are propagated with the given algorithm. During modified sifting no upper limit on the number of **1**-paths is used to stop the swapping operations.

### 5.1.3 Experimental Results

Two experiments were made to demonstrate the difference between the BDDs minimal in the number of **1**-paths and those minimal in size. The first experiment was to enumerate all functions of up to four variables, in the second experiment the benchmark set LGSynth93 has been investigated: The modified sifting algorithm has been compared to Rudell's sifting algorithm. For the modified algorithm also some statistical information has been gathered to validate the efficiency of the technique.

The experiments were carried out on an AMD Athlon 2200+ system with 512 MB of physical memory. The machine was running under Linux. The algorithms were integrated into the CUDD package [Som02]. For the comparison with Rudell's sifting the implementation included in this package was used. CUDD makes use of lower bounds to prune parts of the search space during sifting (see Chapter 4). Also sifting is stopped, as soon as the size of the BDD doubles. No lower bounds or maximal increase are applied during path minimization. Dynamic variable ordering was turned off while building the BDDs, then a single minimization run was carried out.

#### 5.1.3.1 Function Enumeration

For a given function all possible variable orderings have been considered, collecting all BDDs that were minimal in size or number of **1**-paths. Enumerating all functions of two or three variables gives the following result:

LEMMA 5.5 *For all Boolean functions of two or three variables there exists a BDD that is minimal in size* **and** *in the number of* **1***-paths at the same time.*

This is not true for functions of four variables. Comparing the BDDs minimal in size with those minimal in the number of **1**-paths it can be observed that 2.3 percent of the functions do not have a BDD that is minimal in size and the number of **1**-paths at the same time. One of the functions was given in Section 5.1.1.1 already, the corresponding BDDs are shown in Figure 5.1(a) and 5.1(b), respectively. This leads to:

LEMMA 5.6 *There are Boolean functions that do* **not** *have a BDD that is minimal in size and in the number of* **1**-*paths at the same time.*

#### 5.1.3.2 Benchmarks

Due to the propagation of changes towards the terminal node, the minimization of paths is more time consuming than minimizing the size. Not only the descent has to be done but also more memory per node is needed because of the extra information $p_1$, $p_0$, $d_1$ and $d_0$ associated with every node and in addition the stacks are needed. The sifting algorithms were applied after constructing the BDD of a circuit.

In the first experiment statistical information has been collected to judge the quality of the technique. Table 5.1 summarizes the results for some circuits of the LGSynth93 benchmarks the algorithm handled within no more than two hours and no more than 400MByte of memory. The table allows to evaluate the improvements achieved by storing the difference in the number of paths due to a swap and the stacks for propagation of these differences. The last line shows the average percentage of nodes that are visited by the different approaches.

A brute force algorithm would visit the number of nodes listed in column *all nodes*: After each swap the brute force algorithm visits all nodes of the current BDD, even if no change in the number of paths occurred.

Compared to this, *below* gives the sum of nodes that would be visited by an algorithm that only visits nodes below the swapped levels. This is achieved by using additional memory to keep the extra information $p_1$, $p_0$, $d_1$ and $d_0$ for every node.

Finally, column *visited* shows the number of nodes that were visited by the proposed algorithm from Section 5.1.2, i.e. when the stacking schema is used to propagate only from those nodes where a change occurred.

Summarized, the modified algorithm on average only had to visit 3.3% of the nodes a brute force algorithm would visit. Furthermore, for most of the nodes obviously no change in the number of paths had to be propagated. The modified algorithm on average visited only 3.3% of the nodes, while 39.0% of the nodes are below the swapped levels.

In a second run the algorithm was compared to Rudell's sifting algorithm. On some circuits only Rudell's sifting algorithm finished within the given bounds of time and memory, on some benchmarks both al-

*Table 5.1.* Statistical results for the modified sifting algorithm

| name | number of nodes | | |
|---|---|---|---|
| | all nodes | below | visited |
| C432 | 84,639,526 | 41,321,649 | 3,511,386 |
| alu4 | 400,646 | 193,080 | 25,911 |
| apex2 | 19,682,173 | 8,883,647 | 693,229 |
| bigkey | 704,105,565 | 221,187,948 | 88,114 |
| cordic | 416,754 | 161,088 | 13,789 |
| des | 527,491,798 | 199,389,176 | 559,577 |
| dk14 | 5,121 | 1,354 | 509 |
| e64 | 5,974,822 | 2,819,207 | 0 |
| frg1 | 1,636,058 | 624,702 | 67,418 |
| frg2 | 113,200,656 | 31,657,866 | 152,382 |
| i4 | 1052,720,339 | 507,868,694 | 1,678,258 |
| i8 | 66,116,210 | 26,976,058 | 201,564 |
| ldd | 11,420 | 2,255 | 610 |
| pair | 1004,753,819 | 471,731,213 | 2,629,052 |
| phase_decoder | 8,941,404 | 3,475,489 | 64,326 |
| s1238 | 2,900,428 | 967,898 | 71,600 |
| s298 | 238,339 | 103,115 | 32,029 |
| s510 | 1,742,733 | 738,318 | 179,805 |
| scf | 2,927,922 | 1,243,434 | 114,109 |
| too_large | 8,695,072 | 4,464,052 | 331,430 |
| x3 | 28,723,432 | 10,930,792 | 58,500 |
| x4 | 11,671,642 | 3,802,703 | 11,753 |
| Average | 100.0% | 39.0% | 3.3% |

gorithms did not finish. Table 5.2 lists only results for circuits both algorithms were able to cope with.

The table shows for both algorithms the number of nodes and the number of **1**-paths of the BDD, respectively, and the CPU time needed for each circuit. The average factor between the two approaches for corresponding measures is given in the last line. Reducing the number of **1**-paths comes at a higher computational cost. This is due to the non-locality of the swap operation with respect to the number of paths. Another reason is the efficient implementation of sifting in CUDD using lower bounds (see Chapter 4). On the other hand increased computation time often is less important than a good final result. This is especially true for applications like synthesis or test. The size on aver-

*Table 5.2.* Comparison of BDDs resulting from Rudell's and modified sifting

| name | in | out | gates | Rudell | | | **1**-path | | |
|---|---|---|---|---|---|---|---|---|---|
| | | | | size | $P_1$ | time | size | $P_1$ | time |
| C432 | 36 | 7 | 398 | 4,244 | $1.85 \cdot 10^6$ | 1.17 | 19,799 | $1.38 \cdot 10^6$ | 1,700.55 |
| alu4 | 14 | 8 | 2,417 | 599 | 2,410 | <0.01 | 755 | 1,544 | 0.18 |
| apex2 | 39 | 3 | 3,230 | 882 | 114,232 | 0.09 | 2,056 | 8,286 | 9.09 |
| bigkey | 487 | 421 | 3,464 | 3,677 | 8,729 | 0.21 | 2,238 | 7,273 | 2.18 |
| cordic | 23 | 2 | 2,104 | 77 | 34,804 | <0.01 | 121 | 26,330 | 0.06 |
| des | 256 | 245 | 2,764 | 3,143 | 403,487 | 0.16 | 4,788 | 381,927 | 21.17 |
| dk14 | 7 | 8 | 95 | 58 | 95 | <0.01 | 70 | 69 | <0.01 |
| e64 | 65 | 65 | 718 | 157 | 65 | 0.03 | 1,764 | 65 | 0.16 |
| frg1 | 28 | 3 | 206 | 111 | 461 | 0.08 | 154 | 167 | 8.65 |
| frg2 | 143 | 139 | 2,048 | 1,727 | 9,456 | 0.05 | 2,165 | 6,010 | 1.14 |
| i4 | 192 | 6 | 484 | 823 | $1.36 \cdot 10^7$ | 0.36 | 1,783 | $2.30 \cdot 10^6$ | 161.09 |
| i8 | 133 | 81 | 1,304 | 1,336 | 5,683 | 0.05 | 2,235 | 2,230 | 1.57 |
| ldd | 9 | 19 | 126 | 69 | 137 | <0.01 | 79 | 63 | <0.01 |
| my_adder | 33 | 17 | 292 | 243 | $1.24 \cdot 10^6$ | 0.84 | 83 | 655,287 | 4213.40 |
| pair | 173 | 137 | 2,688 | 8,022 | $4.77 \cdot 10^6$ | 0.49 | 9,634 | 106,532 | 24.31 |
| phase_dec. | 59 | 65 | 1,661 | 1,076 | 5,626 | 0.02 | 1,084 | 1,771 | 0.20 |
| s1238 | 33 | 32 | 752 | 636 | 4,168 | 0.01 | 1,179 | 2,621 | 0.33 |
| s510 | 26 | 13 | 222 | 199 | 308 | 0.18 | 208 | 207 | 301.48 |
| scf | 35 | 63 | 694 | 507 | 725 | 0.10 | 603 | 518 | 10.29 |
| too_large | 38 | 3 | 1,152 | 722 | 45,904 | 0.10 | 937 | 4,200 | 2.00 |
| x3 | 135 | 99 | 1,639 | 744 | 2,437 | 0.02 | 691 | 873 | 0.15 |
| x4 | 94 | 71 | 768 | 373 | 1,422 | 0.01 | 765 | 790 | 0.06 |
| Average | | | | 1 | 1 | 1 | 1.92 | 0.54 | 403 |

age increased only by a factor of 1.92 when the modified algorithm was applied. Moreover due to the heuristic approach of sifting the size was reduced for several circuits, even if the number of **1**-paths is minimized (e.g. "my_adder").

The number of **1**-paths was reduced for all the circuits. An average reduction to 54% of the **1**-paths has been achieved. In several cases a factor of 10 or more was gained. For example consider "pair" where the number of **1**-paths was reduced to only 2.2% compared to sifting. Thus, the minimization of the number of **1**-paths instead of the number of nodes can result in significant improvements.

## 5.2 Minimization of Expected Path Length

In this section basic and advanced techniques for BDD minimization with respect to the *expected path length* are presented. The advanced methods are fast heuristic approaches based on sifting and, unlike previous approaches, perform variable swaps with the same time complexity as the original sifting algorithm.

### 5.2.1 Basic Approaches

In [LWHL01], a first algorithm to minimize the EPL for a given single-rooted BDD has been suggested.

The minimization is performed with a window permutation algorithm (see Chapter 4). After each swap of variables, establishing a new variable ordering, the resulting EPL of the restructured BDD must be determined again. Re-calculations need to be done in the restructured part of the BDD covered by the window. Therefore, the window (and with that, also the BDD) is split into an upper and a lower part. Each part is updated with one of two different methods, using formulas derived from Equation (5.7) in Section 2.4.8. Afterwards, this algorithm needs to determine all direct references from nodes in the upper BDD part to nodes in the lower BDD part. This is a significant drawback, as this operation requires touching large parts of the BDD including many levels (see Section 3.1.2.1). The authors also describe a simplified approach which computes an estimation of the EPL, computing only the edges between two adjacent levels. However, this means that a much weaker delay model is used.

In [NMSB03] it has been suggested to use the framework of the sifting algorithm (see Section 2.4.7.2), targeting the EPL of BDDs as the objective function. The authors use the following formula for the expected path length which can be found in [ISM03]. It relates EPL to quantities computed for single BDD nodes:

$$\epsilon(F) = \sum_{i=0}^{|F|-1} P(v_i) \tag{5.5}$$

where $P(v)$ denotes the fraction of all $2^n$ variable assignments such that the resulting path includes node $v$. The global epsilon value for the BDD is determined using the associativity of the sum. For this, it suffices to maintain all changes in the node probabilities as induced locally by each variable swap.

In the next sections an approach will be presented which was published independently of this approach and follows a similar idea. It also directly updates the global EPL while visiting the nodes involved during a variable swap. That way, expensive graph traversals can be avoided. It relates a global change in the quantity in question, i.e. the EPL-value for the BDD, to a local change in the very *same* quantity, i.e. a change in the $\epsilon$-value for a node $v$ currently processed during a swap of levels $i$ and $i+1$.

Let $\Delta_\epsilon$ denote the global change and $\Delta_{\epsilon_v}$ denote the local change, respectively. The basic recurrent equation used in the new method is

$$\Delta_\epsilon = \sum_{\substack{v \text{ is a node} \\ \text{in level } i}} \Delta_{\epsilon_v} \cdot \omega(v), \tag{5.6}$$

where $\omega(v)$ is the weight of node $v$ in the global change.

With that the new approach follows a more general schema which can be used to express other BDD characteristics as well. In particular, the change in the number of **1**-paths that is induced by a variable swap can be described by a similar equation, see Section 5.3.

Moreover, the algorithmic hardness to reduce BDDs with respect to this criteria can be characterized by looking at the respective weights: weights as the above $\omega$ for the EPL-characteristic are *invariant* with respect to changes in the graph structure as caused by a variable swap. Other weights lack this property, e.g. **1**-path reduction based on sifting cannot be performed with a local approach, see Sections 5.1 and 5.3.1.

### 5.2.2 Fast Minimization Approach for Expected Path Length

In this section the latest method to minimize the EPL in BDDs is presented, called EPL-sifting. Like the simple method from [LWHL01] it is based on Rudell's sifting algorithm.

The main result of this section will be the desired *locality* of the variable swap operation in the new method: only the nodes in the adjacent levels affected by a variable swap have to be processed during a swap (due to the re-calculation of values which have become invalid).

With that, a local behavior as in the sifting algorithm has been achieved. It is this locality which makes the new approach fast. In general, it is difficult to obtain locality, e.g. for a sifting modification targeting MPL (MPL-sifting), a local approach is not known. Later in Section 5.2.5, a comparison of the run times of the new method and MPL-sifting is given.

```
eps_on_level(BDD F, int level)
  proc
    for each node v in level level do
      ε(v) := 1 + 1/2(ε(then(v) + ε(else(v)));
    end–for
  end–proc

eps_above_level(BDD F, int level)
  proc
    for i := level to 1 do
      eps_on_level(F, i);
    end–for
  end–proc
```

*Figure 5.7.* Iterative computation of $\epsilon(F)$.

#### 5.2.2.1 Computation of the Expected Path Length

The computation straightforwardly follows two equations that have been introduced in Section 2.4.9.2. These are the equations

$$\epsilon(v) = \begin{cases} 0, & v \in \{\mathbf{1}, \mathbf{0}\} \\ 1 \quad + \quad \text{pr}(\text{var}(v) = 1) \cdot \epsilon(\text{then}(v)), & \text{else} \\ \qquad + \quad \text{pr}(\text{var}(v) = 0) \cdot \epsilon(\text{else}(v)) & \end{cases} \tag{5.7}$$

and

$$\epsilon(F) = \frac{1}{m} \sum_{i=1}^{m} \epsilon(o_i). \tag{5.8}$$

In Figure 5.7, a graph traversing algorithm to compute the $\epsilon$-values for all nodes in a BDD $F$ is given: for a call **eps_above_level**($F$, $n$) the algorithm proceeds bottom-up starting an iteration at the lowest level $n$ onto the highest level 1. On every level the $\epsilon$-values of nodes at the lower levels which already have been computed, are used to compute the current values. To keep the presentation simple, code handling the boundary case ($v \in \{\mathbf{1}, \mathbf{0}\}$) is omitted and equal probability of one and zero assignments for every variable is used for the algorithm in Figure 5.7, i.e. $\text{pr}(x = 0) = \text{pr}(x = 1) = \frac{1}{2}$ for every variable $x$.

EXAMPLE 5.7 *An example for the recursive computation of the $\epsilon$-values is illustrated by the left BDD in Figure 5.8. First, the $\epsilon$-values for the two terminal nodes* **1** *and* **0** *are determined (in both cases, this is the value 0). Next, the $\epsilon$-value for the node labeled with $x_4$ is computed*

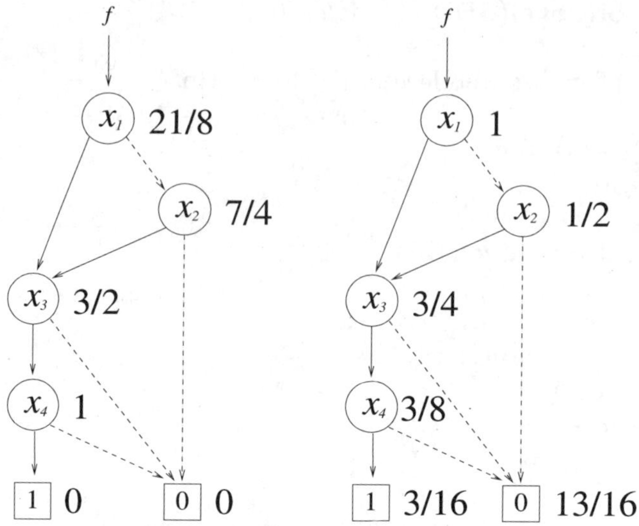


*Figure 5.8.* Recursive computation of the $\epsilon$- and the $\omega$-values.

*as* $1 + \frac{1}{2} \cdot (0 + 0) = 1$ *since both child nodes of that node are terminal nodes. In the next step of the algorithm given in Figure 5.7, the* $\epsilon$*-value of the node labeled with* $x_3$ *then is computed following Equation 5.7 as* $1 + \frac{1}{2} \cdot (1 + 0) = \frac{3}{2}$ *since the* 1*-child is the node labeled with* $x_4$ *and the* 0*-child is the terminal node* **0**. *Next, the* $\epsilon$*-value of the node labeled with* $x_2$ *is computed as* $1 + \frac{1}{2} \cdot (\frac{3}{2} + 0) = 1 + \frac{3}{4} = \frac{7}{4}$*, using the* $\epsilon$*-values of the child nodes, i.e. the node labeled with* $x_3$ *and the terminal node* **0**. *Finally, the* $\epsilon$*-value of the root node labeled with* $x_1$ *is computed using the* $\epsilon$*-values of the child nodes as* $1 + \frac{1}{2} \cdot (\frac{3}{2} + \frac{7}{4}) = 1 + \frac{1}{2} \cdot \frac{13}{4} = 1 + \frac{13}{8} = \frac{21}{8}$.

#### 5.2.2.2 Sifting Algorithm

The basics of the sifting algorithm have already been discussed in Section 2.4.7.2 (see also Figure 2.11). A swapping operation between $\pi(i)$ and $\pi(i+1)$ only changes the graph structure of levels $i$ and $i+1$, leaving all other levels unchanged. This is also true regarding the number of nodes: only the changes in the number of nodes on levels $i$ and $i+1$ contribute to the change of the total BDD size.

Now let the above swap be a step in sifting: if the algorithm given in Figure 5.7 would be incorporated directly into the sifting algorithm, targeting the minimization of EPL instead of BDD size, we would have to re-calculate at least all levels above level $i$ after the swap has been done. This is, because the $\epsilon$-values for nodes on levels $1, \ldots, i-1$ (including output nodes directly influencing the total $\epsilon$-value for the BDD, see

Equation (5.8)) have become invalid since the graph structure below them has changed. This re-calculation would be very time-consuming, yielding a delay minimization routine with a time complexity as high as the one proposed in [LAB98].

#### 5.2.2.3 Keeping Track of Local Changes

To obtain a more efficient algorithm, the new method keeps track of the *local* changes in the $\epsilon$-values of the nodes in the two levels affected by the variable swap. These changes then contribute to a *global* change in the $\epsilon$-value for the BDD. In this, the asymptotic time complexity of the swap operation (see Section 2.4.7.1) is not increased: only the nodes in the affected levels are touched.

The intuition behind Equation (5.6) is the following: both the local changes $\Delta_{\epsilon_v}$ and the node weights $\omega(v)$ can be computed during the variable swap, looking only at nodes which are touched anyway. The local changes induce changes in the $\epsilon$-values of nodes situated at levels $1, \ldots, i-1$ as can easily be seen by Equation (5.7). In particular, also the output nodes are affected. The change in their values induces a change in the global value $\epsilon(F)$ by Equation (5.8). This is the only change of interest.

Equation (5.6) *directly* expresses the impact of the local changes on the global change. In this the method *bypasses* all nodes in levels $1, \ldots, i-1$. As a consequence, only the nodes situated at levels $i$, $i+1$ must be considered. Figures 5.9 and 5.10 sketch the framework for the overall method and for a swap operation following the above idea. Note that in Figure 5.9 in line (3) all node attributes must be set to initial values and the initial global $\epsilon$-value must be computed before sifting starts. While variables are moved up and down by sifting, these moves are recorded. In line (7) the variable is moved back to a best position after this position has been determined. This is done by use of recorded moves.

During sifting, local changes are transferred after each swap to a global change. Then the respective global $\epsilon$-value can be updated, see Figure 5.10. In line (5), the variables $x$ and $y$ are initialized as the variables belonging to the swapped levels ($\pi$ denotes the current variable ordering). For simpler presentation, equal probability of one and zero assignments are assumed for every variable. Line (10) shows the update of the $\epsilon$-value of the re-expressed node $v$ following Equation (5.7). It is straightforward to use the same equation for nodes situated at levels $i+1$, if necessary. Later in Section 5.2.4.1 the schema for the local updates of the involved nodes will be discussed in detail. Lines (9) to (12) show the transfer of the local changes to the global change $\Delta_\epsilon$. If the swap is part of an upward variable movement, in line (15) the $\epsilon$-values of all nodes situated

```
eps_sifting(BDD F, int n)
   proc
      initialize ε- and ω-values of all nodes and ε(F);
      for i := 1 to n do
         use calls to eps_swap to move i-th variable up and
         down until a best position is determined;
         thereby update ε(F) by Δ_ε as returned by
         eps_swap;
         best := sift_back();
         eps_above_level(F, best − 1);
      end–for
   end–proc
```

*Figure 5.9.* Modified sifting algorithm.

```
double eps_swap(BDD F, int x_level, int y_level, enum
direction siftdir)
   proc
      ...
      Δ_ε := 0;
      x := π(x_level); y := π(y_level);
      for each node v with var(v) = x whose represented
      sub-function f_v depends on y do
         re-express v as parent of (possibly newly created)
         child nodes then and else;
         update ε(then), ε(else), ω(then), and ω(else)
         following the local update-schema;
         ε_old := ε(v);
         ε(v) := 1 + 1/2 · (ε(then) + ε(else));
         Δ_ε_v := ε(v) − ε_old;
         Δ_ε := Δ_ε + Δ_ε_v · ω(v);
      end–for
      if siftdir = up then
         eps_on_level(F, x_level − 1);
      end–if
      return Δ_ε;
   end–proc
```

*Figure 5.10.* Swapping two variables.

one level above the swapped levels are restored after the swap. This is necessary to ensure validity of the $\epsilon$-values in subsequent steps as will be discussed in Section 5.2.4.2.

## 5.2.3 Node Weights

Prior to the presentation of update schemas for $\omega$- and $\epsilon$-values appropriate methods to compute both quantities must be known. A recursive method to compute $\epsilon$ has already been given in Figure 5.7, see also the left BDD in the example given in Figure 5.8. This method is used in line (10) of Figure 5.10. In this section the weights $\omega(v)$ for a node $v$ are discussed and it is clarified how to compute them. First, paths from output nodes to $v$ are considered such that the impact of local changes on the change for the output nodes can be expressed. Then an intuitive characterization of the weights is possible. Using this characterization, a recurrent equation is deduced for the weights. With that the node weights can be computed and updated using local operations. A brief comparison to density measures [RS95, RMSS98] is given. Afterwards an invariance property is deduced which helps to avoid unnecessary computations.

### 5.2.3.1 Theory of Weights

For the following, always a shared BDD $F$ representing a Boolean multi-output function $f\colon \mathbf{B}^n \to \mathbf{B}^m$; $f = (f_i)_{1 \leq i \leq m}$ is assumed without further mentioning. The following notation has been introduced in Section 2.4: for an edge $e$ on a path in a BDD, the type of the edge is denoted with $\mathrm{t}(e)$, i.e. we have $\mathrm{t}(e) = 1$ for a 1-edge $e$ and $\mathrm{t}(e) = 0$ for a 0-edge $e$.

LEMMA 5.8 *Let $v$ be a non-terminal BDD node $v$ whose $\epsilon$-value has lately been changed by a value $\Delta\epsilon_v$. Further, assume the structure of the BDD has not changed above this node, i.e. in levels above the* $\mathrm{var}(v)$*-level. Then the $\epsilon$-value of an output node $v'$ changes for every path $p = (v', e_{v'}, v'', e_{v''}, v''', e_{v'''}, \ldots, u, e_u, v)$ by* $\mathrm{pr}(p) \cdot \Delta\epsilon_v$.

**Proof.** We have

$$\mathrm{pr}(p) = \mathrm{pr}\left(\mathrm{var}(v') = \mathrm{t}(e_{v'})\right) \cdot \ldots \cdot \mathrm{pr}(\mathrm{var}(u) = \mathrm{t}(e_u)) \tag{5.9}$$

for the path probability $\mathrm{pr}(p)$ since, for $p$ to be chosen in an evaluation, all variables tested along $p$ must be assigned the according Boolean values. Now the result follows from Equation (5.7): the product of assignment probabilities on the right side of Equation (5.9) is exactly the factor occurring before the term $\epsilon(v)$ in an expression for $\epsilon(v')$, derived with the developed, non-recurrent form of Equation (5.7).

In detail: by repeated application of Equation (5.7) we obtain

$$\begin{aligned}
\epsilon(v') &= \operatorname{pr}(\operatorname{var}(v') = \operatorname{t}(e_{v'})) \cdot \epsilon(v'') + \ldots \\
&= \operatorname{pr}(\operatorname{var}(v') = \operatorname{t}(e_{v'})) \cdot (\operatorname{pr}(\operatorname{var}(v'') = \operatorname{t}(e_{v''})) \cdot \epsilon(v''') + \ldots) \\
&\quad + \ldots \\
&\ldots \\
&= \underbrace{\operatorname{pr}(p)}_{\substack{\text{factor of the}\\ \text{change}}} \cdot \underbrace{\epsilon(v)}_{\substack{\text{subject to}\\ \text{a change } \Delta_{\epsilon_v}}} + \underbrace{\ldots}_{\substack{\text{the term } \epsilon(v)\\ \text{does not occur here}}}
\end{aligned}$$

Note that we ordered the terms on the right sides of the equations such that the terms being multiplicated with $\Delta_{\epsilon_v}$ (and hence, being influenced by this change) are written first. Hence, $\epsilon(v')$ changes by $\operatorname{pr}(p) \cdot \Delta_{\epsilon_v}$. □

In the following, $\operatorname{pr}(p)$ is called the *weight* of $v$ in $v'$ along the considered path $p$ from $v'$ to $v$.

Next, the weight just introduced is formally defined, as well as additional weights which are based on the first weight. For this, the Greek letter $\omega$ will be used with one or two arguments. Formally, for a BDD $F = (\ldots, (V, E), \ldots)$, the weights $\omega$ are functions

$$\omega_F^{(1)}\colon V \to \mathbb{R} \text{ or } \omega_F^{(2)}\colon V \times V \to \mathbb{R},$$

respectively. The BDD $F$ will be given from the context. Thus, to keep the presentation simple, the subscript $F$ normally will be omitted. Of course *all* paths from $v'$ to $v$ must be considered. To compute the total change $\Delta\epsilon_{v'}$, the weights along all paths from $v'$ to $v$ must be summed up. Let

$$\omega(v, v') := \sum_{\substack{p \text{ is a path}\\ \text{from } v' \text{ to } v}} \operatorname{pr}(p) \tag{5.10}$$

denote this total weight of $v$ in $v'$. We have $\Delta\epsilon_{v'} = \Delta\epsilon_v \cdot \omega(v, v')$. By Equation (5.10), $\omega(v, v')$ can also be interpreted as the overall probability of an evaluation reaching $v$ from $v'$. Next, the weight of a non-terminal node $v$ in the change of the global EPL-value is expressed. This change and the weight is denoted $\Delta\epsilon$ and $\omega(v)$, respectively. In analogy to $\omega(v, v')$, the weight of $v$ in another node $v'$, $\omega(v)$ expresses the overall probability of an evaluation reaching $v$ from an output node, i.e.

$$\omega(v) := \frac{1}{m} \sum_{i=1}^{m} \omega(o_i, v)$$

where $o_i$ is the output node representing $f_i$.

Note that by Equation (5.10) the terms $\omega(o_i, v)$ still depend on an enumeration of all paths from $v'$ to $v$. Fortunately, it is also possible to express $\omega(v)$ independently of path enumerations.

LEMMA 5.9 *Let $v, v'$ be non-terminal nodes in a BDD $F$. Then the weight of $v$ in the change of $\epsilon(F)$ can be expressed as*

$$\omega(v) = \frac{k}{m} + \sum_{\substack{v \text{ is } b\text{-child} \\ \text{of } v',\ b \in \mathbb{B}}} \operatorname{pr}\left(\operatorname{var}(v') = b\right) \cdot \omega(v') \qquad (5.11)$$

*where $k$ is the number of single-output functions $f_i$ which are represented by $v$ (if any).*

**Proof.** The proof is by induction on the stage of the transitive closure of the edge relation (constructed as intermediate steps while building the transitive closure of the graph), combined with a case analysis. Let $\mathcal{S}$: $\{F \mid F \text{ is a BDD}\} \times (\mathbb{N} \cup \{0\}) \to 2^{\{v \in V \mid (\ldots,(V,E),\ldots) \text{ is a BDD}\}}$ be defined as

$$\mathcal{S}(F, k) = \begin{cases} O, & k = 0 \\ \{v \in V \mid (u, v) \in E \text{ and } u \in \mathcal{S}(F, k-1)\}, & k > 0 \end{cases}$$

where $F = (\ldots, (V, E), O)$. With $\mathcal{S}$, different stages of the transitive closure of $E$ are obtained for different second arguments $k$: first, $\mathcal{S}(F, 0)$ yields the set of output nodes in $F$. These are the starting point of any evaluation with $F$, defining the first stage. Then $\mathcal{S}(F, 1)$ yields the image of these nodes, i.e. the set of their child nodes. This corresponds to testing the first variable on an arbitrary evaluation path and following the respective edges. In general, $\mathcal{S}(F, k)$ ($k > 0$) is the $k$-th stage in the transitive closure of the outlined image relation. $\mathcal{S}$ eventually reaches a fixed point, i.e. $\mathcal{S}(F, k) = \mathcal{S}(F, k+1) = \emptyset$, corresponding to the fact that the terminal nodes in $\{\mathbf{1}, \mathbf{0}\}$ eventually must be reached by every evaluation (the set of child nodes of the terminal nodes is empty). First we assert that $\forall v \in V$: $\exists k \geq 0$: $v \in \mathcal{S}(F, k)$: this is due to all component graphs of $F$ being connected.

The proof now is by induction on $k$, the second argument of $\mathcal{S}$. First, let $k = 0$. Let $v \in \mathcal{S}(F, 0)$. Then $v$ must be an output node without parent nodes, i.e. $v$ is a root node representing an output function.

By Equation (5.8) it is easily seen that the contribution of an output node to the global change $\Delta\epsilon$ must be $\frac{k}{m}$ (this also expresses the probability of an evaluation starting at output node $v$). This is correctly expressed by Equation (5.11): the sum term

$$\sum_{\substack{v \text{ is } b\text{-child} \\ \text{of } v',\, b \in \mathrm{B}}} \mathrm{pr}\,(\mathrm{var}(v') = b) \cdot \omega(v')$$

vanishes since $v$ has no parent nodes.

Now assume the claim is shown for all $v \in \mathcal{S}(F,k)$. Let $v \in \mathcal{S}(F,k+1)$. We distinguish two cases:

**Case a):**

$v \notin O$, i.e. $v$ is no output node. By assumption, $v$ must have at least one parent node and $v$ itself is not an output node. Thus to reach $v$ from any output node, first a parent node $v'$ must be reached. Hence $v' \in \mathcal{S}(F,k)$ by definition of $\mathcal{S}$. Moreover, if $v$ is a $b$-child of $v'$, $\mathrm{var}(v')$ must be assigned $b$ to reach $v$ from $v'$. By induction hypothesis, the weight of $v'$ in the change of $\epsilon(F)$ (or, in other words, the overall probability of an evaluation reaching $v'$ from an output node) can be expressed as $\omega(v')$ as defined in Equation (5.11). Hence, the product of the according probabilities, $\omega(v')$ and $\mathrm{pr}(\mathrm{var}(v') = b)$, expresses the probability of an evaluation reaching $v$ via $v'$, starting at an output node. Since these probabilities are accumulated over all parent nodes, in fact the total probability of an evaluation reaching $v$ from an output node is obtained. This is equivalent to the weight of $v$ in the change of $\epsilon(F)$.

**Case b):**

$v \in O$, i.e. $v$ is an output node. In Equation (5.11), $\frac{k}{m}$ is added to the sum term. Hence, with the result for $k = 0$ and the result from Case a) it is clear that the case of $v$ being an inner output node is also handled correctly. □

### 5.2.3.2 Computation of Weights

An example for the recursive computation of the $\omega$-values is illustrated by the right BDD in Figure 5.8. Note that the sum of $\omega(\mathbf{1})$ and $\omega(\mathbf{0})$ must always equal 1, as all evaluations either end at **1** or end at **0**, i.e. the sum of the according probabilities must be 1. Next, an algorithm to compute the $\omega$-values is given. The code in Figure 5.11 assumes that all values $\omega(v)$ are initially set to zero. Starting from the uppermost

```
calc_omega(BDD F, int n)
  proc
    for each node v representing an output function do
      ω(v) := ω(v) + 1/m;
    end–for
    for i := 1 to n do
      for each node v in level i do
        increase ω(then(v)) and ω(else(v)) by ½ω(v);
      end–for
    end–for
  end–proc
```

*Figure 5.11.* Iterative computation of the $\omega$-values.

level, the $\omega$-values are propagated down from parent nodes to its child nodes which are residing on lower levels (see Figure 5.11). Again, code for handling the boundary case is omitted and equal probability of one and zero assignments is assumed for every variable to keep the presentation simple. In Section 5.2.4 we will see that this algorithm is only necessary for the initialization of the weights. In particular, it is *not* necessary to use this algorithm after each variable swap and hence the local behavior of the new approach is preserved. Note that the top-down algorithm to compute $\omega$ has a run time which is linear in the BDD size. In contrast, node characteristics like density measures [RS95, RMSS98] are computed bottom-up. They measure the size of a cofactors ONSET while the weights described here are not related to the ONSET. However, node weights similar to the weights here have been used in [RMSS98] to express density changes.

Inspecting Equation (5.11) one might expect a change of $\omega(v)$ to influence $\omega$-changes for all descendants of $v$. This would be a potential problem if a sifting modification targeting the EPL must update $\omega$-values after a swap. Fortunately, the situation is much simpler which is shown in the next section.

#### 5.2.3.3 Invariance of Weights

The next results are crucial for the desired *locality* of the new approach. The probabilistic interpretation of the weights allows to conclude a useful invariance property. A first step is to turn attention to probabilities of variable assignments instead of path probabilities.

LEMMA 5.10 *Let $F$ be a BDD representing a Boolean multi-output function $f$. Let $v$ be a node on level $k$ in $F$ representing the Boolean function $f_v$ (a cofactor of $f$ with respect to the first $k-1$ variables in the ordering). Further, let $b = (b_1, \ldots, b_{k-1})$. Then the weight $\omega(v)$ can be expressed as follows:*

$$\omega(v) = \sum_{\substack{b \text{ in } \mathbf{B}^{k-1} \\ \text{with } f_b = f_v}} \mathrm{pr}(x_1 = b_1) \cdot \ldots \cdot \mathrm{pr}(x_{k-1} = b_{k-1}) \qquad (5.12)$$

REMARK 2 *The following notation was introduced in Section 2.4: if an evaluation of the BDD with respect to the (partial) assignment $a = (a_1, a_2, \ldots, a_k)$ $(k \leq n)$ stops at a (non-terminal) node $v$, $v$ represents the cofactor $f_{x_1=a_1,\ldots,x_k=a_k}$ (due to the semantics of BDDs). This cofactor is also simply denoted $f_a$.*

**Proof of Lemma 5.10.** In the previous section it was clarified that $\omega(v)$ expresses the probability of an evaluation reaching $v$ from an output node.

Equation (5.12) expresses the same probability using the fact that every evaluation reaching a node defines a cofactor which is functionally equivalent to the function represented by the node: previously, the probabilities of an evaluation traversing certain *paths* have been considered. Now, as the key observation, the probability of *assignments* $b$, such that

$$f_b = f_v \qquad (5.13)$$

are considered instead (note that here a variable $x_i$ is not necessarily tested along some path in the BDD since simply assignments to the first $k-1$ variables of the ordering are considered). By the BDD semantics of Section 2.4, an evaluation of $F$ with respect to an assignment must reach $v$ iff it is satisfying the condition in Equation (5.13). Hence, the total probability of an evaluation reaching $v$ can be expressed as the sum of the probabilities of all assignments which satisfy the condition in Equation (5.13). The probability of such an assignment is the product of the respective probabilities for the single variable bindings as given by the assignment. This yields Equation (5.12). □

Still Equation (5.12) depends on the BDD's graph structure, but now only because the cofactor $f_b$ must equal the function represented by node $v$. In constrast to Equation (5.11), Equation (5.12) expresses $\omega(v)$ as a *function* of $f$, $f_v$, and the level of $v$ only. Since a BDD representing a fixed $f$ is considered, the following result follows immediately:

THEOREM 5.11 *Let $F$ be a BDD representing a Boolean function $f$ and let $v$ be a node in $F$. Fixed probabilities are assumed for the variable assignments to values in* **B**. *Then, if a) the function represented by $v$ and b) the number of the $v$-level are preserved with respect to a change in the variable ordering (and thus, in the graph structure) of the BDD, the value $\omega(v)$ also does not change, i.e. $\omega(v)$ is* invariant *with respect to this change.*

The background is that the swap operation of the sifting algorithm re-expresses nodes $v$ in the upper level of the two levels affected by the swap. Yet, a node $v$ is expressed by a new tuple $(\text{var}(v), \text{then}(v), \text{else}(v))$, but in this case this does not change the function represented (as the ordering of variables as well has changed, see Section 2.4.6). Even more important, all nodes in the unaffected levels clearly represent the same function and stay in their respective levels after the swap. But then $\omega(v)$, as a weight depending on these two quantities only, cannot change either.

## 5.2.4 Update Schema

Equations (5.7) and (5.11) are appropriate for a propagation of changes in the $\epsilon$- and $\omega$-values of child nodes to their parents or vice versa, respectively. This allows to establish valid values for all nodes in the two levels involved by a variable swap by local operations, i.e. only the nodes in these two levels must be touched. For a presentation of the final method to minimize the expected path length in BDDs the framework shown in Figure 5.10 still needs refinement. For this purpose, a local update schema will be presented in Section 5.2.4.1, describing all actions taken depending on the respective situation.

A problem that has not been addressed so far is that value changes in levels $i$, $i+1$ may imply changes in levels different from $i$ and $i+1$. As subsequent swaps base their computations on these values, one has to take care that only valid values are used in these swaps. For this reason in Section 5.2.4.2 a general update-schema, i.e. a schema for updates after all swaps for a particular variable have been carried out, is given. It describes which levels are affected and what values need to be re-calculated in different situations.

### 5.2.4.1 Local Update Schema

Let us consider the situation of a variable swap: each swap of two adjacent levels $i$ and $i+1 =: j$ changes the local graph structure of the BDD in these two levels, leaving all other levels unchanged, e.g. see Figure 5.12. Again equal probability of one and zero assignments for

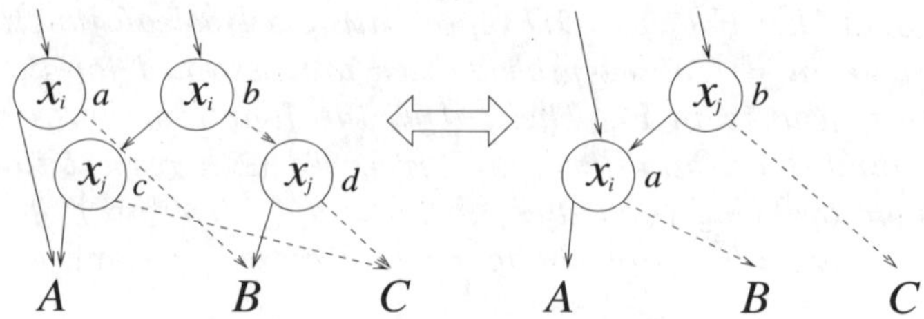


*Figure 5.12.* Swap of two BDD levels $i$ and $i+1$ $(=: j)$.

every variable are assumed. Before the swap, $x_i$ resides in level $i$, $x_j$ in level $j$. The local structural changes cause changes in the $\epsilon$- and $\omega$-values of nodes in the affected levels. E.g. see node $a$ in Figure 5.12: after the swap of levels $i$ and $j$: node $a$ has become a 1-child of node $b$. Thus $\frac{1}{2}\cdot\omega(b)$ is added to the old $\omega$-value of $a$, updating $\omega(a)$ following Equation (5.11). Node $b$ has new child nodes after the swap. Hence $\epsilon(b)$ is re-calculated using the $\epsilon$-values of the new child nodes (node $a$ and the root node of sub-graph $C$) following Equation (5.7).

Also note that $b$ still represents the same function and still is on the same level after the swap. Consequently, $\omega(b)$ remains unchanged by Theorem 5.11. The same invariance holds for all nodes outside the swapped levels. The change in the $\epsilon$-value of nodes $v$ in level $i$ is the value $\Delta\epsilon_v$ described in Section 5.2.2.3.

Next all cases to consider during a swap are given in detail with the respective actions taken.

**Changes of $\epsilon$-Values.** First, the changes of $\epsilon$-values are considered.

**Case a):**

The $\epsilon$-values of nodes which were in levels $i+1,\ldots,n$ *before* the swap do not change since the levels below them are not touched by the swap and a node's $\epsilon$-value depends on the $\epsilon$-values of the child nodes only (see Equation (5.7)).

**Case b):**

Depending on the functional equivalence of certain nodes on levels below $i+1$, some nodes might been newly created on level $i+1$ during the swap (for details see Section 2.4.7.1). Hence, an initial $\epsilon$-value for these nodes must be computed. Since the $\epsilon$-values of all levels below these new nodes are still valid (see Case a)), this can be done with only two lookups[1] of the $\epsilon$-values of the 1-child and 0-child. E.g. in Figure 5.12, going from right to left, node $b$ is re-expressed as parent of nodes $c$ and $d$, both of which are newly created nodes. The initial value for $\epsilon(c)$ is computed from the $\epsilon$-values of the roots of sub-graphs $A$ and $C$, the one for node $d$ from the $\epsilon$-values of the roots of sub-graphs $B$ and $C$.

**Case c):**

Nodes in level $i$ need re-calculation of the $\epsilon$-value during the swap if they have been re-arranged to connect to newly created or different child nodes. The values of the child nodes have already been computed in Case b) or a), hence re-calculation involves only two lookups. E.g. in Figure 5.12, going from right to left, $\epsilon(b)$ is updated as $1+\frac{1}{2}\cdot(\epsilon(c)+\epsilon(d))$, following Equation (5.7).

The change in the $\epsilon$-value of these nodes $v$ in level $i$ is the value $\Delta\epsilon_v$ described in Section 5.2.2.3. These are the only changes which need to be transferred to the global $\epsilon$-value. The changes of $\epsilon$ for the child nodes in Case b) do not need to be transferred since these changes already have been transferred to the parent nodes in level $i$.

**Case d):**

The $\epsilon$-values of all nodes in levels $1, \ldots, i-1$ have become invalid. This is because their $\epsilon$-values depend on child nodes which might have been changed. Later, discussing the general update schema, it turns out that re-calculating all of them after every swap can be avoided.

**Changes of $\omega$-Values.** Now the changes in the weights $\omega$ are considered.

**Case a):**

The $\omega$-values of all nodes in levels $1, \ldots, i-1$ remain valid by the result of Theorem 5.11 as these levels are not touched and the function represented by the nodes is not changed during the swap operation.

**Case b):**

By Theorem 5.11, a node $v$ in level $i$ does not need re-calculation of the $\omega$-weight either. The node is re-expressed representing the same function and still is situated at the same level.

---

[1] These lookups are implemented with standard hashing techniques. In the experiments, these operations showed the expected quasi-constant behavior.

*Table 5.3.* Levels affected by the swap

| changed value | affected level |
|---|---|
| $\epsilon$-value | the levels with numbers $i+1, i, i-1, \ldots, 0$ |
| $\omega$-value | the level with number $i+1$ |

**Case c):**

Again, let us consider Figure 5.12, going from right to left. On level $i+1$ new nodes $c$ and $d$ are created during the swap. An initial $\omega$-value must be computed for these nodes. They are established as child nodes of node $b$. The weight of node $b$ is already valid (see Case b). The weight of a newly created node is initialized with the sum of weight contributions of the parents: for each parent node on level $i$ with weight $\omega$, this contribution is $\frac{1}{2} \cdot \omega$, following Equation (5.11). In the example, $\omega(c)$ as well as $\omega(d)$ are initialized with $\frac{1}{2} \cdot \omega(b)$.

Nodes that, after the swap, become a new child node of a node in level $i$, may have existed before (e.g., see node $a$ in Figure 5.12 when going from left to right). The function they represent is preserved (this follows from the Shannon decomposition (see Theorem 2.9) since neither the variable tested nor the child nodes are changed), but the level on which these nodes reside changes.

Hence, Theorem 5.11 does not apply in this case and the $\omega$-value of $v$ has to be updated: After the swap, let $v$ be a re-expressed node on level $i$ which is parent of a node $v'$. Following Equation (5.11), $\frac{1}{2} \cdot \omega(v)$ is added to $\omega(v')$ (in the example, $\frac{1}{2} \cdot \omega(b)$ is added to $\omega(a)$).

**Case d):**

The $\omega$-values of nodes which are in levels $i+2, \ldots, n$ need not be updated after the swap by Theorem 5.11 since the level at which they are situated as well as the function represented by them remains unchanged: this is due to locality of the swap operation.

### 5.2.4.2 General Update-Schema

Table 5.3 summarizes which levels can contain invalid nodes that need re-calculation of the $\epsilon$- and $\omega$-value (during or after a swap of the levels $i$ and $i+1$). Updating the nodes with invalid $\omega$-values does not change the asymptotical complexity of the swap operation since the nodes in level $i+1$ must be touched anyway.

At a first glance, the updates of the $\epsilon$-values seem to be a problem as all levels above the affected levels can contain invalid nodes. Fortunately, the desired local behavior can still be achieved since the method does not always require that all values are valid.

To see this, first recall that the sifting algorithm first moves a variable up and down (or vice versa), then back to its best position. First assume a variable $x$ being currently at level $i$ is moved *down* and consider a node $v$ with $\text{var}(v) = x$. To determine the correct $\epsilon$-value of $v$, nothing but the validity of the $\epsilon$-values of the child nodes of $v$ is required. Hence, re-calculation of the $\epsilon$-value of the invalid nodes in levels $1, \ldots, i-1$ *is not necessary* during the whole sequence of swaps moving down variable $x$.

Suppose now the variable $x$ has reached the bottommost level $n-1$. Now the $\epsilon$-values of nodes above the last two BDD levels, i.e. on levels $1, \ldots, n-2$ are invalid. Next, sifting moves variable $x$ upwards. Assume variable $x$, on its way up, is situated at level $i$. The next swap performed will be one of the levels $i-1$ and $i$. Again, $\Delta\epsilon_v$ is computed for a node $v$ on level $i-1$. But then the *old* $\epsilon$-value of $v$ *before* this next swap must be valid, otherwise the value of $\Delta\epsilon_v$ is incorrect (see lines (9), (11) in Figure 5.10).

As a remedy, as $x$ is moved up again, it suffices to re-calculate after each swap the $\epsilon$-values for the nodes on the level directly above the two swapped levels. For this reason, whenever the direction of movement is upwards, **eps_on_level**($F$, $i-1$) is called after a swap of levels $i$ and $i+1$ (see line (15) in Figure 5.10).

This leads to the summary of levels subject to value updates after a swap of levels $i$ and $i+1$ as shown in Table 5.4. Different from Table 5.3, now only a *constant* number of levels has to be maintained with every swap. Thus, locality of the desired sifting modification remains. After having tried all positions by moving a variable $x_i$ up and down, $x_i$ is moved back to a best position seen in the previous movements. Let this position be on level $k$. To reestablish proper $\epsilon$-values for subsequent movements of the other variables, all $\epsilon$-values of nodes above level $k$ are re-calculated with a call to **eps_above_level**($F$, $k-1$) (see line (8) in Figure 5.9). This is an operation touching many levels, i.e. a large part of the whole BDD. But this only has to be done *once* after *all* swaps to determine the best position for a variable have been performed and hence does not lead to a higher asymptotical time complexity than that of the original sifting algorithm.

To summarize: a modification of the original sifting algorithm has been found which targets the expected path length in BDDs. This algorithm has the same asymptotical time complexity as the original sifting algo-

*Table 5.4.* Levels subject to updates after/during the swap

| sifting direction | affected levels wrt... | |
|---|---|---|
| | $\epsilon$ | $\omega$ |
| moving a variable up | $i+1, i, i-1$ | $i+1$ |
| moving a variable down | $i, i+1$ | $i+1$ |

rithm as a sophisticated schema ensures that no traversal on the whole graph of the BDD is necessary when performing a variable swap. This schema consists of keeping track of local changes in the levels affected by the swap and then directly transferring these changes to a change in the global EPL-value.

### 5.2.5 Experimental Results

In this section, experimental results are presented. All algorithms have been applied to circuits of the LGSynth93 benchmark set [Col93]. The tested methods include EPL-sifting and maximal path length sifting (MPL-sifting), i.e. a sifting modification targeting the maximal path length in BDDs. For a comparison also the standard sifting algorithm has been applied. A weaker form of MPL-sifting with a simplified objective function has been used in [SB00]. As no approach with local behavior is known for MPL-sifting, this algorithm has high run times. All algorithms have been integrated into the CUDD package [Som02] and were tested in the same system environment. A system with an Athlon CPU running at 1.4 GHz with a main memory of 1.5 GByte has been used for the experiments. All methods use BDD size as second criterion in case of ties of the first criterion. In the tests, equal probabilities of one and zero assignments for every variable have been used. This seems to be the most realistic approach if specific assumptions about these probabilities cannot be made. In this, the tests were performed with the same assumptions as in the approach of [LWHL01] which also used equal probabilities.

In a first series of experiments MPL- and EPL-sifting have been compared. Therefore both sifting approaches have been applied to the test-cases given in Table 5.5. In the first column the name of the function is given. Column *in* gives the number of inputs of a function. Column *size* shows the initial size (given as number of BDD nodes) of the BDD representing the function. In columns *MPL* and *EPL* the maximal path length and the expected path length of the initial BDD for a function are given.

The results of the experiments are given in Table 5.6. In the first column again the name of the function is given. The next three columns *size*, *MPL* and *EPL* give the size, the maximal path length and the expected path length for the BDD after applying the MPL-sifting approach. The next column *time* shows the run time of the minimization by MPL-sifting in CPU seconds. In the next four columns the same quantities size, MPL, EPL and run time are shown for EPL-sifting.

EPL-sifting is slower than size-driven sifting. This degradation is by a small factor which remains the same regardless of the number of input variables of the considered function. This is due to the constant overhead caused by hash table lookups which are necessary to access the $\omega$- and $\epsilon$-values associated with BDD nodes. In this, EPL-sifting is of the same asymptotical complexity as classical sifting. As the results show, EPL-sifting achieves the same reduction in MPL as MPL-sifting: both EPL- and MPL-sifting improve the initial MPL by 8.3% in average and the improvement can be up to 48.9% (e.g., see *dalu*). There is no significant difference in the results of the two methods, i.e. choosing EPL-sifting instead of MPL-sifting preserves high quality of the results. However, applying EPL-sifting instead of MPL-sifting has a great advantage: EPL-sifting is the *fastest* delay-driven minimization approach preserving high quality of the results. The total time (average) speed-up factor of EPL-sifting compared to MPL-sifting is 8.25 and can be up to two orders of magnitude (e.g., see *i7*).

Moreover, EPL-sifting also yields the best starting point for critical path analysis, [JKCMS97, BMP97, LAB98] as the resulting EPL-values (corresponding to the average gate delay) are much better than for the other methods. EPL-sifting reduces the initial EPL by 29.3% on average. The improvement can be up to 63.4%, especially for larger instances (e.g., see *dalu*). EPL-sifting achieves an EPL which is better by 14.5% on average compared to the EPL yielded by MPL-sifting. The improvement over MPL-sifting can be up to 40.1% (e.g., see *x4*).

In a second series of experiments, also "classical" size-driven sifting has been applied to the set of test-cases. The accumulated results and total run times of the two series of experiments are given in Table 5.7. The first column *criterion* gives the optimization criterion targeted by the used method (here, the entry "initial" means the BDDs before any method has been applied). Columns two to four each state the sum of a BDD characteristic over the BDDs for all test-cases: in column two, this characteristic is BDD size, in column three it is MPL and in column four it is EPL. The last column states the total run time for the whole

*Table 5.5.* Initial values for the test-cases

| name | in | size | MPL | EPL |
|---|---|---|---|---|
| apex6 | 135 | 2759 | 21 | 3.56 |
| apex7 | 49 | 1659 | 24 | 5.03 |
| b9 | 41 | 177 | 13 | 3.17 |
| c1355 | 41 | 43869 | 41 | 30.76 |
| c3540 | 50 | 223227 | 30 | 10.94 |
| c499 | 41 | 39377 | 41 | 29.65 |
| c5315 | 178 | 5247 | 47 | 4.00 |
| c880 | 60 | 15544 | 42 | 5.65 |
| cht | 47 | 149 | 5 | 2.71 |
| dalu | 75 | 12946 | 47 | 16.05 |
| example2 | 85 | 468 | 16 | 2.54 |
| frg2 | 143 | 2230 | 22 | 3.00 |
| i3 | 132 | 132 | 32 | 4.46 |
| i4 | 192 | 420 | 47 | 6.56 |
| i5 | 133 | 311 | 19 | 2.50 |
| i6 | 138 | 412 | 4 | 3.10 |
| i7 | 199 | 504 | 4 | 3.25 |
| i8 | 133 | 3980 | 16 | 5.42 |
| i9 | 88 | 2270 | 12 | 5.48 |
| k2 | 45 | 2012 | 24 | 5.03 |
| pair | 173 | 13845 | 49 | 6.08 |
| rot | 107 | 8322 | 57 | 4.27 |
| s5378 | 199 | 5218 | 41 | 3.70 |
| s641 | 54 | 1351 | 27 | 3.69 |
| s713 | 54 | 1351 | 27 | 3.69 |
| s838.1 | 66 | 244 | 55 | 4.08 |
| x1 | 51 | 1296 | 23 | 3.88 |
| x3 | 135 | 945 | 20 | 3.00 |
| x4 | 94 | 890 | 15 | 3.79 |

test-suite (the "-" for row *initial* means nothing has been done here at all).

All methods also improve BDD size significantly. In this, EPL-sifting yields only 7.3% more size than MPL-sifting (and classical size sifting) on average. This can allow higher circuit speed at only low area overhead.

*Table 5.6.* Comparison of MPL- and EPL-sifting

| name | MPL sifting | | | | EPL sifting | | | |
|---|---|---|---|---|---|---|---|---|
| | size | MPL | EPL | time | size | MPL | EPL | time |
| apex6 | 639 | 20 | 2.37 | 6.59s | 662 | 20 | 2.33 | 0.17s |
| apex7 | 310 | 19 | 2.59 | 0.54s | 274 | 19 | 2.25 | 0.05s |
| b9 | 107 | 13 | 3.04 | 0.18s | 125 | 13 | 2.65 | 0.02s |
| c1355 | 30326 | 41 | 26.70 | 126.11s | 39201 | 41 | 20.82 | 64.97s |
| c3540 | 73319 | 30 | 10.66 | 1171.64s | 46228 | 30 | 9.81 | 131.01s |
| c499 | 30459 | 41 | 27.41 | 369.72s | 38465 | 41 | 20.34 | 54.71s |
| c5315 | 2611 | 47 | 3.89 | 127.28s | 3137 | 47 | 4.07 | 1.44s |
| c880 | 4865 | 41 | 5.27 | 25.09s | 9883 | 41 | 4.82 | 3.49s |
| cht | 89 | 4 | 2.63 | 0.17s | 124 | 4 | 2.06 | 0.02s |
| dalu | 1216 | 24 | 5.88 | 24.73s | 1006 | 24 | 4.94 | 3.13s |
| example2 | 298 | 14 | 2.71 | 1.28s | 393 | 14 | 2.18 | 0.06s |
| frg2 | 1434 | 20 | 2.42 | 15.98s | 1648 | 20 | 2.32 | 0.37s |
| i3 | 132 | 32 | 4.46 | 4.57s | 132 | 32 | 4.46 | 0.08s |
| i4 | 300 | 47 | 4.78 | 15.52s | 300 | 47 | 4.38 | 0.21s |
| i5 | 133 | 19 | 1.98 | 5.32s | 133 | 19 | 1.98 | 0.08s |
| i6 | 208 | 4 | 3.24 | 4.69s | 274 | 4 | 3.05 | 0.07s |
| i7 | 367 | 4 | 3.35 | 15.10s | 434 | 4 | 3.18 | 0.13s |
| i8 | 1678 | 13 | 3.49 | 25.91s | 2495 | 13 | 3.40 | 0.81s |
| i9 | 1659 | 12 | 5.00 | 5.48s | 1821 | 10 | 4.96 | 0.21s |
| k2 | 1355 | 24 | 4.03 | 1.23s | 1438 | 24 | 4.01 | 0.14s |
| pair | 5857 | 49 | 4.13 | 197.76s | 8775 | 49 | 3.76 | 4.40s |
| rot | 6374 | 57 | 4.05 | 48.82s | 15062 | 59 | 3.08 | 7.69s |
| s5378 | 2489 | 33 | 3.43 | 65.50s | 5975 | 34 | 3.11 | 1.43s |
| s641 | 628 | 25 | 3.14 | 0.91s | 724 | 26 | 2.65 | 0.11s |
| s713 | 628 | 25 | 3.14 | 0.92s | 724 | 26 | 2.65 | 0.10s |
| s838.1 | 298 | 38 | 3.37 | 0.98s | 625 | 35 | 2.92 | 0.10s |
| x1 | 487 | 22 | 2.80 | 0.75s | 603 | 22 | 2.67 | 0.07s |
| x3 | 612 | 20 | 2.36 | 7.14s | 669 | 20 | 2.34 | 0.14s |
| x4 | 530 | 15 | 3.99 | 2.46s | 512 | 15 | 2.39 | 0.07s |

## 5.3 Minimization of Average Path Length

This section discusses the minimization of the *average path length* in BDDs. As a by-product it is clarified how other criteria for BDD optimality can be expressed by recurrent equations following the general

*Table 5.7.* Size, MPL, EPL, and run time sums for different criteria

| criterion | $\sum$ size | $\sum$ MPL | $\sum$ EPL | total time |
|---|---|---|---|---|
| initial | 391155 | 821 | 189.04 | – |
| EPL | 181842 | 753 | 133.58 | 275.30s |
| MPL | 169408 | 753 | 156.31 | 2272.37s |
| size | 168914 | 777 | 162.85 | 47.30s |

outline of Equation (5.6). In this, a *unifying view* of sifting modifications is applied, that targets objective functions different from diagram size: global changes of the objective function in question are induced by local changes of the nodes in the topmost level of the two BDD levels involved in the variable swap.

As a first example consider the minimization of **1**-paths in BDDs which has been studied in Section 5.1. Let $\Delta_{\mathbf{1}}$ denote the global change in the number of **1**-paths after a variable swap affecting levels $i$ and $i+1$. Further, let (i) $\Delta_{1_v}$ and (ii) $\Delta_{0_v}$ denote the local changes in the number of (i) regular and (ii) complemented paths from node $v$ to **1**. Then the respective weights in the global change are $p_1(v)$, the number of regular paths from output nodes to $v$ and $p_0(v)$, the number of complemented paths from output nodes to $v$. This results in the following equation:

$$\Delta_1 = \sum_{\substack{v \text{ is a node} \\ \text{in level } i}} (\Delta_{1_v} \cdot p_1(v) + \Delta_{0_v} \cdot p_0(v))$$

Minimization of the **1**-paths in BDDs using the sifting algorithm can yield higher run times than classical, size-driven sifing (see Section 5.1). One reason is that BDDs minimized with respect to **1**-paths can be larger in size than BDDs yielded by classical sifting. Another important reason is the absence of an invariance result corresponding to Theorem 5.11 for the weights $p_1(v)$ and $p_0(v)$. Because a change in these weights for one node $v$ can cause changes in the weights of many descendants, the weight changes must be propagated down to **1** with every swap (see Section 5.1).

A classification of criteria for BDD optimality in terms of algorithmic hardness is possible by means of the invariance property of the respective weights (or its absence, respectively). This classifies EPL-sifting to be on the side of "easy" problems and **1**-path sifting (as well as other problems, e.g. the minimization of the sum of the lengths of the paths) to appear on

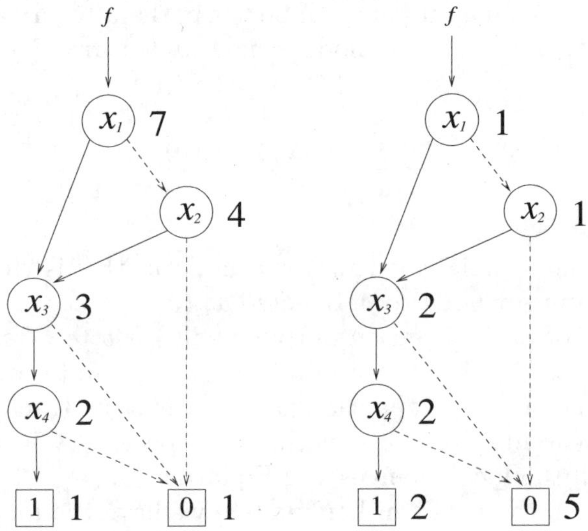


*Figure 5.13.* Recursive computation of $\alpha$- and $\omega_\alpha$-values.

the "hard" side. Next, this unifying view will be used in the presentation of an algorithm to compute the average path length in a BDD.

### 5.3.1 Applying the Unifying View

First, the task to minimize BDDs with respect to the number of paths is reconsidered. Thereby a unifying view helps to understand the algorithmic hardness.

**Number of Paths.** An example of the recursive computation of the $\alpha$-values is illustrated by the left BDD in Figure 5.13. Equation (2.2) is used for computing the number of paths in a BDD in time proportional to the BDD size: similar to the algorithm given in Figure 5.7, every node is visited only once.

A naive incorporation of this strategy into the sifting algorithm would be to simply re-calculate $\alpha(F)$ after each swap. But this involves a traversal of the complete graph of the BDD $F$, resulting in high run times.

Next it is examined whether a *local* method for the computation of the number of paths can be obtained by transferring the idea of EPL-sifting.

By keeping track of all changes in the values $\alpha(v)$, denoted $\Delta_{\alpha_v}$, the change of the global value $\alpha(F)$, i.e. $\Delta_\alpha$, must be computed. Let $\Delta_\alpha$

denote the global change in the number of paths after a variable swap affecting levels $i$ and $i+1$. A node weight $\omega_\alpha(v)$ must be found for a node $v$ such that

$$\Delta_\alpha = \sum_{\substack{v \text{ is a node} \\ \text{in level } i}} \Delta_{\alpha_v} \cdot \omega_\alpha(v). \tag{5.14}$$

In fact such a node weight exists: for a node $v$, this is the number of paths from an output node to $v$, denoted $\omega_\alpha(v)$.

The number of paths from an output node via $v$ to a terminal node is simply $\alpha(v) \cdot \omega_\alpha(v)$, i.e. the product of ingoing and outgoing paths. The local change in $\alpha(v)$ is the change which must be transferred to the global change (by multiplying it with the respective weight $\omega_\alpha(v)$ and summing up all the local changes, see Equation (5.14)).

Analogously to the result in Lemma 5.9, we have the following result for the weight $\omega_\alpha(v)$.

LEMMA 5.12 *Let $v$ be a non-terminal node in a BDD $F$. Then the weight of $v$ in the change of $\alpha(F)$ can be expressed as*

$$\omega_\alpha(v) = k + \sum_{\substack{v \text{ is child} \\ \text{of } v'}} \omega_\alpha(v') \tag{5.15}$$

*where $k$ is the number of single-output functions $f_i$ which are represented by $v$ (if any).*

With $k$ again we account for the fact that one output node may represent several single-output functions.

An example for the recursive computation of the $\omega_\alpha$-values is illustrated by the right BDD in Figure 5.13. Note that the sum of $\omega_\alpha(\mathbf{1})$ and $\omega_\alpha(\mathbf{0})$ is 7 and equals $\alpha(F)$, as denoted at the root node of the left BDD.

Using Equation (5.15) it is straightforward to derive a graph traversing algorithm computing $\omega_\alpha(v)$ similar to the one given in Figure 5.11. (This is left to the interested reader.)

The outlined approach has local behavior iff the weights $\omega_\alpha(v)$ respect an invariance property similar to Theorem 5.11. Unfortunately, this is *not* the case: in the terminology of Section 5.1, it is $\omega_\alpha(v) = p_1(v)+p_0(v)$. In the same section it was described how a change in $p_1(v)$ ($p_0(v)$) affects the $p_1$-value ($p_0$-value) of all nodes on a path from $v$ to $\mathbf{1}$.

The same unifying view helps to understand the algorithmic hardness of computing the sum of the lengths of the paths in a BDD $F$, $\lambda(F)$.

```
lambda_on_level(BDD F, int level)
   proc
      for each node v in level level do
         α(v) := α(then(v)) + α(else(v));
         λ(v) :=
         λ(then(v)) + α(then(v)) + λ(else(v)) + α(else(v));
      end–for
   end–proc

lambda_above_level(BDD F, int level)
   proc
      for i := level to 1 do
         lambda_on_level(F, i);
      end–for
   end–proc
```

*Figure 5.14.* Iterative computation of $\lambda(F)$.

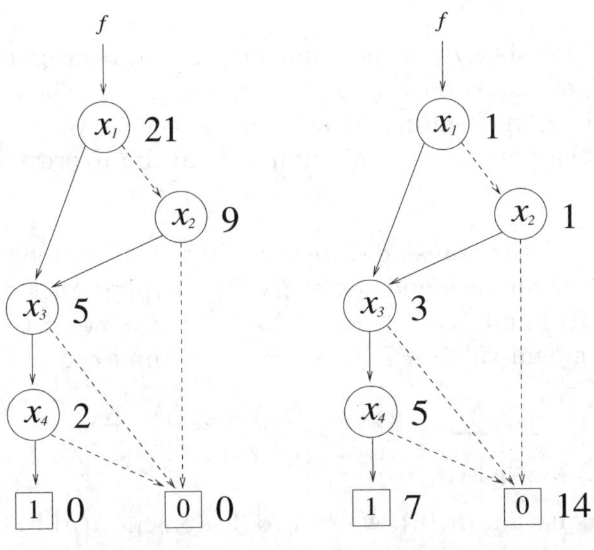


*Figure 5.15.* Recursive computation of $\lambda$- and $\omega_\lambda$-values.

**Sum of the Lengths of the Paths.** For a BDD $F$, $\lambda(F)$ can be computed by a graph traversal following Equation (2.7), see Figure 5.14. To keep the presentation simple, code for the handling of the boundary case ($v \in \{\mathbf{1}, \mathbf{0}\}$) is omitted. An example of the recursive computation of the $\lambda$-values is illustrated by the left BDD in Figure 5.15. Again it would be inefficient to call the algorithm given in Figure 5.14 with every variable swap made in the sifting algorithm.

Before the unified view can be applied to examine whether a method with local behavior is available, the following result is needed. Let the sum of the lengths of all paths from an output node to a node $v$ be denoted $\omega_\lambda(v)$.

LEMMA 5.13 *Let $v$ be a node in a BDD. The sum of the lengths of all paths from an output node via $v$ to a terminal node can be expressed as*

$$\omega_\lambda(v) \cdot \alpha(v) + \omega_\alpha(v) \cdot (\lambda(v) - 1). \tag{5.16}$$

**Proof.** The sum of the lengths of all paths from an output node via $v$ to a terminal node can be expressed as the sum of two numbers $n_1$ and $n_2$: the first number $n_1$ is the sum of the numbers of nodes on the "first half" of each path $p$ via $v$, i.e. the number of nodes on a sub-path of each such path $p$, starting at an output node and ending at $v$. This number can be expressed as $\omega_\lambda(v) \cdot \alpha(v)$: there are $\alpha(v)$ paths emerging from $v$, each of which can be combined with one of the paths going into $v$. That is, each path emerging from $v$ contributes with $\omega_\lambda(v)$ nodes to $n_1$.

The second number $n_2$ is the sum of the numbers of nodes on the "second half" of nodes on a sub-path of each path via $v$, starting at a child node of $v$ and ending at a terminal node. With an argument analogous to the one for $n_1$, this number can be expressed as $\omega_\alpha(v) \cdot (\lambda(v) - 1)$. □

Now a schema for the transfer of local changes to the global change can be given. Let $v$ again be a node in level $i$. By keeping track of all changes in the values $\alpha(v)$ and $\lambda(v)$, denoted $\Delta_{\alpha_v}$ and $\Delta_{\omega_v}$, we can compute the change of the global value $\lambda(F)$, i.e. we can compute $\Delta_\lambda$:

$$\Delta_\lambda = \sum_{\substack{v \text{ is a node} \\ \text{in level } i}} \omega_\lambda(v) \cdot \Delta_{\alpha_v} + \omega_\alpha(v) \cdot \Delta_{\lambda_v}. \tag{5.17}$$

Note that in Equation (5.16), we had to decrement $\lambda(v)$ by one in order not to count node $v$ several times. In the above Equation (5.17) this is not necessary since only a relative *change* in the $\lambda$-value is considered.

Before examining the question whether this is an approach with local behavior, it should be ensured that an efficient algorithm for the initial computation of the weights exists. This algorithm is based on the next result.

In analogy to the result of Lemma 5.12 and of Lemma 5.9, we have the following result for the weight $\omega_\lambda(v)$.

```
calc_omega_lambda(BDD F, int n)
    proc
        for each node v representing an output function do
            ω_α(v) := ω_α(v) + 1;
        end–for
        for i := 1 to n do
            for each node v in level i do
                increase ω_α(then(v)) and ω_α(else(v)) by ω_α(v);
                increase ω_λ(then(v)) and ω_λ(else(v)) by ω_λ(v) +
                ω_α(v);
            end–for
        end–for
    end–proc
```

*Figure 5.16.* Iterative computation of the $\omega_\lambda$-values.

LEMMA 5.14 *Let $v$ be a non-terminal node in a BDD $F$. Then the sum of the lengths of all paths from an output node to $v$ can be expressed as*

$$\omega_\lambda(v) = k + \sum_{\substack{v \text{ is child} \\ \text{of } v'}} \omega_\lambda(v') + \omega_\alpha(v') \tag{5.18}$$

*where $k$ is the number of single-output functions $f_i$ which are represented by $v$ (if any).*

An example for the recursive computation of the $\omega_\lambda$-values is illustrated by the right BDD in Figure 5.15. Note that the sum of $\omega_\lambda(\mathbf{1})$ and $\omega_\lambda(\mathbf{0})$ is 21 and equals $\lambda(F)$, as denoted at the root node of the left BDD. This is because $\lambda(F)$ and $\omega_\lambda(\mathbf{1})+\omega_\lambda(\mathbf{0})$ (or simply $\omega_\lambda(\mathbf{1})$ if we consider BDDs with CEs) are expressions summing up the lengths of the very same paths by definition. Following Equation (5.18), it is straightforward to give a graph traversing algorithm computing $\omega_\lambda(v)$ similar to the one given in Figure 5.11. This algorithm operates level-wise, i.e. all predecessors of a node $v$ have been visited before visiting $v$ itself. The code assumes that all values $\omega_\lambda(v)$ and $\omega_\alpha(v)$ are initially set to zero and is given in Figure 5.16.

The outlined approach following the idea of EPL-sifting has local behavior iff the weights respect an invariance result corresponding to Theorem 5.11. Again, this is *not* the case since weights $\omega_\alpha(v)$ are used which suffer from the absence of such a property. Moreover, the weight $\omega_\lambda(v)$ relies on the weight $\omega_\alpha(v)$ by definition. Hence we have the same influence of a change in $\omega_\lambda(v)$ on the nodes occurring on paths that emerge from $v$.

### 5.3.2 Algorithm

In the previous section a method to minimize the sum of the lengths of the paths in BDDs has been derived by transferring the idea of EPL-sifting. Thereby a unified view of BDD optimization problems has been applied.

Unfortunately, the obtained method does *not* show the local behavior of the EPL-sifting method. The reason for this has been identified as the absence of invariance properties corresponding to Theorem 5.11 for the outlined approach. Consequently, the following is done:

**Number of Paths.** Instead of transferring the local changes $\Delta_{\alpha_v}$ to $\Delta_\alpha$, the algorithm keeps track of all the changes of $\omega_\alpha(v)$. These changes are propagated down to the terminal nodes (or to **1** if BDDs with CEs are used). Thereby the technique described in Section 5.1.2.3 is used. Then $\alpha(F)$, the number of paths from an output node to a terminal node in $F$, equals $\omega_\alpha(\mathbf{1}) + \omega_\alpha(\mathbf{0})$ (or simply $\omega_\alpha(\mathbf{1})$ if BDDs with CEs are used) which, by definition, expresses the very same number of paths.

Note that this exactly describes the algorithm given in Section 5.1, but this time as a consequence of the unifying view.

**Sum of the Length of the Paths.** Since the updates of the weights $\omega_\lambda$ also cannot be avoided, there is no reason not to use them directly: $\lambda(F)$, the sum of lengths of the paths in a BDD $F$, equals $\omega_\lambda(\mathbf{1}) + \omega_\lambda(\mathbf{0})$ (or simply $\omega_\lambda(\mathbf{1})$ if we consider BDDs with CEs) since the two expressions sum up the lengths of the very same paths by definition. Hence, instead of transferring the local changes $\Delta_{\lambda_v}$ to $\Delta_\lambda$, the algorithm keeps track of all changes of $\omega_\lambda(v)$. These changes are propagated down to the terminal nodes (or only to **1** if BDDs with CEs are used), using the following technique.

In Section 5.1 it has been described how the changes of the weights can be efficiently propogated down to the terminal nodes. This technique makes use of a levelized stack and only those nodes are considered during propagation where a change in the number of paths may have occured.

Downward propagation of the weights $\omega_\lambda$ is very similar, except that two more assignments are used for every child $w$ of a node $v$ that is retrieved from the topmost stack.

$$d_\lambda(w) := d_\lambda(w) + d_\lambda(v) + d_\alpha(v) \tag{5.19}$$

$$\omega_\lambda(w) := \omega_\lambda(w) + d_\lambda(v) + d_\alpha(v) \tag{5.20}$$

These two new assignments (5.19) and (5.20) properly follow Equations (5.15) and (5.18) since the propagation is done level-wise, i.e. all predecessors of a node $v$ are already visited before visiting $v$ itself.

Now it is straightforward to give the desired algorithm for minimizing the average path length.

The algorithm sketched in Figure 5.16 initially computes $\omega_\alpha(\mathbf{1})$, $\omega_\alpha(\mathbf{0})$ as well as $\omega_\lambda(\mathbf{1})$, $\omega_\lambda(\mathbf{0})$ (or $\omega_\alpha(\mathbf{1})$ and $\omega_\lambda(\mathbf{1})$ in case of BDDs with CEs). During sifting, variables are moved up and down, using local variable swaps. After each swap, the propagation routine as outlined above reestablishes proper values of the $\omega_\alpha$- and $\omega_\lambda$-weights of the terminal node(s).

As has already been explained before, for a BDD $F$ without CEs we have

$$\begin{aligned} \lambda(F) &= \omega_\lambda(\mathbf{1}) + \omega_\lambda(\mathbf{0}), \\ \alpha(F) &= \omega_\alpha(\mathbf{1}) + \omega_\alpha(\mathbf{0}). \end{aligned}$$

Following Equation (2.6), the new approach to reduce the target criterion APL by use of sifting recomputes $\overline{\lambda}(F)$ in every step simply as

$$\overline{\lambda}(F) = \frac{\omega_\lambda(\mathbf{1}) + \omega_\lambda(\mathbf{0})}{\omega_\alpha(\mathbf{1}) + \omega_\alpha(\mathbf{0})}.$$

After the currently considered variable has been moved back to a best position, a call **calc_omega_lambda**($F$) restores proper weights for all nodes in the BDD.

This finishes the description of a sifting modification targeting the objective function $\overline{\lambda}(F)$ for a BDD $F$.

### 5.3.3 Experimental Results

In this section, experimental results with two different versions of the new approach to reduce the average path length in BDDs are presented. A preliminary version and the final algorithm are applied to test-cases of the LGSynth93 benchmark set [Col93]. Again, in case of ties of the APL criterion, BDD size has been used as second criterion.

In the first algorithm, called “brute force”, the propagation technique as described in Section 5.1.2.3 is not used. The second algorithm is called “final” and makes use of the propagation technique.

Both algorithms have been integrated into the CUDD package [Som02] and were tested in the same system environment. A system with an Athlon CPU running at 1.4 GHz with a main memory of 1.5 GByte has been used for the experiments.

In a series of experiments, the run times of the two versions of the algorithm have been compared and the BDD sizes and APL-values resulting from applying APL-sifting to the set of test-cases have been computed. Therefore both algorithms have been applied to the test-cases given in

Table 5.8. In the first column the name of the function is given. Column *in* gives the number of inputs of a function. Column *initial size* shows the initial size (given as the number of BDD nodes) of the BDD representing the function. In column *final size* the size of the BDD resulting from the application of APL-sifting is given. Since both algorithms implement the same method (more or less efficiently), these results are the same for both algorithms. Column *initial APL* states the average path length of the initial BDD for a function whereas column *final APL* again states that value after application of APL-sifting. Again, these final APL-values are the same for the different implementations of the same approach.

As the results show, APL-sifting yields significant reductions in the APL of a BDD. The improvement is up to 54.6% (see *dalu*). On average, the improvement is 22.9%. However, APL-sifting does not always improve the size of the BDDs and can yield large increases in BDD size (e.g., see *i3*). For one test-case, *c5315*, the size of the BDD resulting from APL-sifting even exceeded the node limit of the system (the final APL given in this case is that of the last BDD that the system was able to build). This contrasts to the experiences with EPL-sifting (see Section 5.2.5) where also the BDD size has been improved in most of the cases and without exception results of only moderate BDD size are obtained. Table 5.9 shows the run times for the different algorithms. In the first column the name of the function is given. The two sub-columns of column *time*, i.e. columns *brute force*, and *final* give the run times in CPU seconds for the according algorithms.

As expected, algorithm "brute force" yields the highest run times. Using the propagation technique of Section 5.1.2.3 instead of a brute force method which has to touch every node below the levels affected by the swap, large speed-ups are achieved. Compared to algorithm "brute force", algorithm "final" can be faster by a factor of two orders of magnitude (e.g., see *s5378*).[2]

## 5.4 Summary

In this chapter, several alternative criteria for BDD optimality which are different from the classical criterion, BDD size, have been studied. These new criteria are motivated by several applications in VLSI CAD.

New efficient approaches for the reduction of BDDs with respect to these criteria have been suggested. They are based on the most success-

[2]Note that the test-case *c5315* has been excluded from the statistics since the size of the BDD resulting from APL-sifting exceeded the node limit of the system in this case. The numbers given in brackets show the time when the node limit (10 mio.) was reached.

*Table 5.8.* Initial values for the test-cases

| name | in | initial size | final size | initial APL | final APL |
|---|---|---|---|---|---|
| apex6 | 135 | 2759 | 1082 | 15.38 | 7.77 |
| apex7 | 49 | 1659 | 612 | 14.06 | 8.77 |
| b9 | 41 | 177 | 213 | 7.57 | 7.12 |
| c1355 | 41 | 43869 | 41768 | 38.23 | 37.35 |
| c3540 | 50 | 223227 | 279863 | 26.77 | 24.50 |
| c499 | 41 | 39377 | 40488 | 38.22 | 37.35 |
| c5315 | 178 | 5247 | >10 mio. | 46.75 | 41.99 |
| c880 | 60 | 15544 | 21527 | 29.84 | 24.57 |
| cht | 47 | 149 | 120 | 3.09 | 2.82 |
| dalu | 75 | 12946 | 2985 | 27.62 | 12.53 |
| example2 | 85 | 468 | 522 | 8.12 | 6.73 |
| frg2 | 143 | 2230 | 2917 | 14.13 | 10.57 |
| i3 | 132 | 132 | 393216 | 23.50 | 21.01 |
| i4 | 192 | 420 | 3603 | 31.77 | 28.21 |
| i5 | 133 | 311 | 494 | 7.35 | 6.03 |
| i6 | 138 | 412 | 274 | 3.32 | 3.31 |
| i7 | 199 | 504 | 366 | 3.43 | 3.36 |
| i8 | 133 | 3980 | 4091 | 11.98 | 8.10 |
| i9 | 88 | 2270 | 2059 | 8.17 | 6.85 |
| k2 | 45 | 2012 | 1690 | 13.53 | 8.95 |
| pair | 173 | 13845 | 37836 | 28.05 | 20.24 |
| rot | 107 | 8322 | 97734 | 38.96 | 32.55 |
| s5378 | 199 | 5218 | 12566 | 35.82 | 21.74 |
| s641 | 54 | 1351 | 1125 | 14.84 | 10.51 |
| s713 | 54 | 1351 | 1125 | 14.84 | 10.51 |
| s838.1 | 66 | 244 | 779 | 44.89 | 20.69 |
| x1 | 51 | 1296 | 831 | 13.06 | 11.20 |
| x3 | 135 | 945 | 1068 | 13.52 | 8.15 |
| x4 | 94 | 890 | 720 | 8.39 | 7.65 |

ful approach to dynamic reordering known so far, the sifting algorithm of Rudell (see Section 2.4.7.2).

First, the *number of paths* in BDDs has been studied. Several applications directly benefit from a smaller number of paths in a BDD. Among these are the minimization of DSOPs, CNF generation and test of BDD circuits. Theoretical results show that an exponential differ-

*Table 5.9.* Run times for APL-sifting in two stages of development

| name | time | |
|---|---|---|
| | brute force | final |
| apex6 | 31.08s | 0.58s |
| apex7 | 1.49s | 0.13s |
| b9 | 0.51s | 0.04s |
| c1355 | 415s | 85.4s |
| c3540 | 4485s | 826s |
| c499 | 334s | 77.7s |
| c5315 | – | (16323s) |
| c880 | 262s | 34.9s |
| cht | 0.47s | 0.04s |
| dalu | 127s | 9.01s |
| example2 | 3.93s | 0.11s |
| frg2 | 84s | 1.29s |
| i3 | 14172s | 151.3s |
| i4 | 189s | 2.83s |
| i5 | 13.34s | 0.24s |
| i6 | 11.43s | 0.18s |
| i7 | 31.03s | 0.47s |
| i8 | 150s | 2.68s |
| i9 | 21.19s | 0.62s |
| k2 | 3.51s | 0.42s |
| pair | 3769s | 41.4s |
| rot | 5823s | 123s |
| s5378 | 761s | 7.64s |
| s641 | 2.55s | 0.25s |
| s713 | 2.58s | 0.25s |
| s838.1 | 3.89s | 0.70s |
| x1 | 2.13s | 0.19s |
| x3 | 29.01s | 0.50s |
| x4 | 6.69s | 0.18s |

ence in the number of paths can result from different variable orderings, while the number of nodes is linear. Also the variable ordering leading to the minimal number of nodes can be different from the variable ordering leading to the minimal number of paths. An efficient algorithm to reduce the number of paths has been introduced and compared to an algorithm reducing the size. Experiments showed that the overhead in size is moderate while the number of paths can be reduced by several orders of magnitude for some benchmarks.

Next, an efficient technique to optimize BDDs with respect to the expected path length has been presented. This criterion measures the evaluation time for functional simulation. It also models average gate delay assuming a unit delay model. The method is based on sifting and uses a new sophisticated approach to keep track of local changes and their impact on the global change during a variable swap. Thus it achieves small run times, staying within the time complexity of the original sifting algorithm.

Experimental results are reported demonstrating that the proposed technique also results in BDDs optimized for the synthesis of fast multiplexor circuits. This is achieved with a significant smaller run time than classical approaches.

The two criteria which have been studied, the number of paths and the expected path length in BDDs, have been compared and characterized applying a unifying view. With the help of this general framework, methods to optimize BDDs with respect to other criteria, like the sum of the lengths of the paths and the average path length in BDDs, can be derived. Experiments showed the feasibility of the obtained approach to minimize the average path length in BDDs.

# Chapter 6

# RELATION BETWEEN SAT AND BDDS

State-of-the-art verification tools are based on efficient operations on Boolean formulas. Besides BDDs also *Boolean Satisfiability* (SAT) solvers are frequently used. In this chapter we study the relation between SAT and BDDs. Formal equivalence checking is exemplarily considered as an application.

SAT has received increased attention as a promising technique for *Automatic Test Pattern Generation* (ATPG) [Lar92, SBSV92], model checking [BCCZ99], and equivalence checking. Virtually all SAT techniques rely on the use of the *Davis-Putnam procedure* (DP) [DP60] and the *Davis-Logeman-Loveland procedure* [DLL62] to explore the search tree. In the following we do not distinguish the two, but simply refer to the DP procedure. If there is a pattern that differentiates the circuits under verification, then DP will eventually find it or prove that the SAT formula is unsatisfiable. Numerous techniques have been proposed to reduce the search tree. Some of these techniques such as iterated global implications [SBSV92] and recursive learning [MSG99] are applied as a preprocessing step, while others are applied during the course of the application of the DP procedure. For example, clause recording [MS98] and cache-based backtracking [PCK99] are used to avoid conflicts. Non-chronological backtracking [MSS96] and efficient implementations [MMZ$^+$01] further improve the performance. An alternative research trend focuses on identifying variable ordering techniques that minimize the number of backtracks executed by the DP procedure to find a satisfying assignment [GN02]. However, especially for unsatisfiable formulas often run time is the limiting resource.

On the other hand, BDDs have been traditionally used to solve the equivalence checking problem due to their canonicity. However, it is this requirement for canonicity that makes BDDs inefficient in representing certain classes of functions. For example, integer multipliers have displayed exponential memory requirements for any variable ordering [Bry91].

There has been increased interest in techniques that integrate SAT and BDDs to reduce the time and space needed to solve the equivalence checking problem [BS98, PK00, RS01]. Though it has been noted that SAT and BDDs represent the same entity [PCK99], there is little understanding of the relation between the two procedures and how the techniques of one domain can be utilized in the other. First approaches considered preprocessing a problem using BDDs to reduce the search space for the SAT solver [CNQ03, GGW+03]. Another idea was to improve the reasoning capabilities of a SAT solver by using BDDs [GYA+01].

The presentation in this chapter follows the approach from [RDO02]. The relations between the search tree of a SAT solver and the number of paths in a BDD are studied. Equivalence checking is considered to illustrate this relation. The equivalence checking problem is viewed as a search in the decision trees of the two circuits for a path that leads to the terminal node **1** (**0**) in one but leads to the terminal node **0** (**1**) in the other. From this perspective, it would be desirable to decrease the number of paths, thus reducing the number of backtracks and time needed to solve the problem. A dynamic variable ordering technique to run the DP procedure is proposed. The technique is geared towards minimization of the number of backtracks needed to prove the *unsatisfiability* of the CNF formula that results from proving the equivalence of two circuits. This method is compared with the greedy variable ordering of TEGUS [SBSV92]. Experimental results verify that the proposed approach results in a dramatic decrease in the number of backtracks and time needed to solve the problem.

The organization of the chapter is as follows. The DP procedure is reviewed in Section 6.1 to make the book self-contained. Section 6.2 presents the theoretical foundations for the relation between the DP procedure and BDDs. Section 6.3 proposes a dynamic BDD-based variable ordering technique for the DP procedure. Experimental results are given in Section 6.4, followed up by summary in Section 6.5.

## 6.1 Davis-Putnam Procedure

A CNF formula $\rho$ is a set of clauses where each clause is the disjunction of a number of literals where a literal is a variable or its negation. Since each logic gate can be represented by a number of clauses as shown in [Lar92], a CNF formula of a logic circuit is the conjunction of the CNF

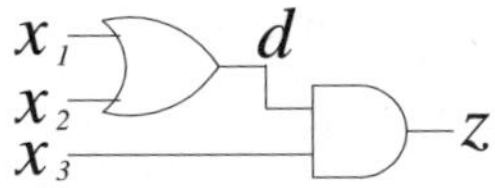


*Figure 6.1.* Circuit for Examples 6.1 and 6.6.

formulas of all the gates. A CNF formula is satisfiable if at least one set of assignments to the variables of the formula makes it evaluate to 1.

EXAMPLE 6.1 *Consider the circuit in Figure 6.1. The OR-gate is modeled by following three clauses:*

$$(\overline{x}_1 + d) \cdot (\overline{x}_2 + d) \cdot (x_1 + x_2 + \overline{d})$$

*This CNF is only satisfied for an assignment that does not lead to a conflict at the OR-gate, i.e. that can occur in the actual circuit. In the same way the AND-gate is modeled by clauses. The conjunction of the CNFs for the OR-gate and the AND-gate yields a CNF $\rho$ that models the circuit:*

$$\rho = (\overline{x}_1 + d) \cdot (\overline{x}_2 + d) \cdot (x_1 + x_2 + \overline{d}) \cdot (x_3 + \overline{z}) \cdot (d + \overline{z}) \cdot (\overline{d} + \overline{x}_3 + z)$$

Virtually all SAT solvers use the DP procedure [DP60, DLL62] in their core in order to find a satisfying assignment for the formula or conversely to prove the formula unsatisfiable. The DP procedure performs a backtracking depth-first search in the space of all truth assignments to find a satisfying assignment for the CNF formula. The performance of backtracking is greatly improved by employing unit clause propagation: whenever a unit clause arises, the variable occurring in that clause is assigned the truth-value that satisfies the clause. The formula is thereupon simplified which may lead to new unit clauses. Figure 6.2 outlines the DP procedure. The procedure returns 1 in case the CNF formula is satisfiable, 0 otherwise. At first a literal to split on is chosen (line 14). If no literal is left, the CNF is already satisfied and 1 is returned (lines 15-17). Otherwise both possible assignments for the literal are assigned (lines 18-22 and 24-28). For each assignment the procedure is recursively called to solve the sub-problems (lines 19 and 25, respectively). In case both sub-problems are unsatisfiable 0 is returned.

## 6.2 On the Relation between DP Procedure and BDDs

In this section we study the relation between the DP procedure and the BDD representation of the same circuit. We first formalize the various properties of the CNF formulas generated from multi-level combinational logic circuits and show how these properties allow a measure of

```
assign(sat_formula ρ, literal ν)
   proc
      ν := 1 in ρ;
      simplify ρ;
      apply unit clause propagation;
      if ρ has an empty clause then
         return 0;
      else
         return 1;
      end–if
   end–proc

DP(sat_formula ρ)
   proc
      choose literal ν to split on;
      if ν = NULL then
         return 1;
      end–if
      if assign(ρ, ν) then
         if DP(ρ) then
            return 1;
         end–if
      end–if
      undo ν assignment;
      if assign(ρ, ν̄) then
         if DP(ρ) then
            return 1;
         end–if
      end–if
      return 0;
   end–proc
```

*Figure 6.2.* DP procedure.

flexibility in the search for a satisfying assignment to the formula. Furthermore, the relation between the number of backtracks obtained using the DP procedure and the number of paths in the corresponding BDD is proven. This relation allows the calculation of optimal lower bounds for the number of backtracks needed to prove the equivalence of two equivalent circuits. We start by introducing some notation to provide a concise basis for the formalization of the derived results.

Let $\rho$ be a CNF formula, and $V(\rho)$ denote the set of variables that $\rho$ depends on. A clause $c_i \in \rho$ is satisfied if there is some assignment to its literals such that the disjunction of the literals evaluates to 1. The CNF formula $\rho$ is satisfied if there is a truth assignment to $V(\rho)$ such that every clause $c_i \in \rho$ is satisfied.

Let $CCNF$ denote the set of CNF formulas generated from multi-level combinational logic circuits. Let $\rho \in CCNF$ represent a multi-level combinational circuit $C$, and let $f$ be the underlying Boolean function of $C$. Since the logic value of all the internal and output signals of $C$ can be derived from the values of the primary inputs, we consider the support of $f$ to be the primary inputs only. The set $X_n$ represents the $n$ primary inputs of $\rho$. In addition, we introduce the variable $z$ to reference the primary output of the circuit. The relation between $\rho$, $X_n$ and $V(\rho)$ is given by the next lemma.

LEMMA 6.2 *If $\rho \in CCNF$, then it is possible to find a set of variables $X_n \subset V(\rho)$ such that $\rho$ can be satisfied by only splitting on the variables of $X_n$ in the DP procedure.*

**Proof.** Since the internal nodes of a multi-level logic circuit are functions of the primary inputs, it is possible to determine their assignment given some assignment on the set of primary input variables, $X_n$. □

Lemma 6.2 enables a reduction of the search space from one in terms of all the variables of the circuit nodes to one in terms of the primary inputs. A similar reduction was exploited by the PODEM algorithm for test pattern generation [Goe81]. In the DP procedure this reduces the backtracking to be only in terms of the primary inputs. In a typical run of the DP procedure, a sequence of support variables will be assigned truth variables during the search of a satisfying assignment to the whole CNF formula. This subset of currently assigned primary inputs shall be denoted by $S$, and their truth assignments by $A_S$. The restricted Boolean function that results from the application of $A_S$ to $\rho$ shall be denoted by $f_{A_S}$.

In order to avoid fruitless searches, the DP procedure should avoid assigning truth-values to variables that can make no contribution to the satisfiability of the formula. This notion is captured by the following definition.

DEFINITION 6.3 *If $A_S$ is the truth assignment of a set $S \subseteq X_n$ and $\nu \in X_n$ but $\nu \notin S$, then $\nu$ is said to be redundant under $A_S$ if*

$$f_{A_S}|_{\nu=0} = f_{A_S}|_{\nu=1}.$$

Definition 6.3 simply states that if the restricted Boolean function $f_{A_S}$ is insensitive to the change in $\nu$, then there is no point in assigning $\nu$ a truth value. This notion allows us to tailor down the satisfiability of CCNF formulas as given by the following theorem.

THEOREM 6.4 *A CNF formula $\rho \in CCNF$ is satisfied under a truth assignment $A_S$ of a set $S \subseteq X_n$ if $\forall \nu \in (X_n \setminus S) : f_{A_S}|_{\nu=0} = f_{A_S}|_{\nu=1}$.*

**Proof.** If $\forall \nu \in (X_n \setminus S) : f_{A_S}|_{\nu=0} = f_{A_S}|_{\nu=1}$, then no truth assignment to any element that belongs to $(X_n \setminus S)$ can change the function $f(A_S)$. Since all variables in $S$ are assigned truth values under $A_S$ and $(X_n \setminus S) \cup S = X_n$, then it follows from Lemma 6.2 that $\rho$ is satisfied. □

Theorem 6.4 proves that if all the remaining unassigned primary inputs are redundant, then there is no need for additional variable assignments since the $CCNF$ formula is satisfiable. Another important property of $CCNF$ formulas is that there is no point in assigning additional variables if the primary output has already been assigned a truth value. This is proved in the next theorem.

THEOREM 6.5 *If $\rho \in CCNF$ and $z$ is assigned a truth value under a truth assignment $A_S$ of a set $S \subseteq X_n$, then $\rho$ is satisfied.*

**Proof.** If $z$ is assigned a truth value, then $\forall \nu \in (X_n \setminus S) : f_{A_S}|_{\nu=0} = f_{A_S}|_{\nu=1}$. This is true since none of the variables that belongs to $(X_n \setminus S)$ can change the primary output $z$ corresponding to $f$. Thus, it follows from Theorem 6.4 that $\rho$ is satisfied. □

Theorem 6.5 defines our notion of satisfiability for *Circuit Satisfiability* (CSAT). In addition, it opens the possibility for the primary output variable $z$ to be assigned a truth-value while there exist clauses that are not evaluated to 1 and have some remaining unassigned literals. Example 6.6 illustrates such a case.

EXAMPLE 6.6 *The CNF formula of the circuit in Figure 6.1 is $\rho = (\overline{x}_1 + d) \cdot (\overline{x}_2 + d) \cdot (x_1 + x_2 + \overline{d}) \cdot (x_3 + \overline{z}) \cdot (d + \overline{z}) \cdot (\overline{d} + \overline{x}_3 + z)$. In this example, the circuit has three primary inputs $X_3 = \{x_1, x_2, x_3\}$ and the primary output $z$. Under the partial assignment $A_S = \{x_3 = 0\}$, where $S = \{x_3\}$, $z$ is assigned the truth value 0 and $\rho = (\overline{x}_1 + d) \cdot (\overline{x}_2 + d) \cdot (x_1 + x_2 + \overline{d})$. We notice that under $A_S$ both $f_{A_S}|_{x_1=0} \oplus f_{A_S}|_{x_1=1}$ and $f_{A_S}|_{x_2=0} \oplus f_{A_S}|_{x_2=1}$ equal zero since the function output value is already determined. Furthermore, there exists a truth assignment to the set of variables $X_n \setminus S$ that satisfies the remaining clauses. This truth assignment is however of no interest since the primary output has* *already been assigned a truth-value. As an illustration for Definition 6.3, if $A_S = \{x_1 = 1\}$ where $S = \{x_1\}$ then $\rho = (x_3 + \overline{z}) \cdot (\overline{x}_3 + z)$ and $f_{A_S}|_{x_2=0} = f_{A_S}|_{x_2=1}$, yet $z$ is not assigned a truth value. Thus, $\rho$ is not satisfied but $x_2$ is redundant under $A_S$.*

We notice that the Deletion Rule from Section 2.4.3 for BDD reduction is equivalent to Definition 6.3 in the CSAT context and thus tracing a path from the root to the **1**-terminal or **0**-terminal in the BDD of circuit $C$ is equivalent to finding a satisfying assignment to the CNF formula $\rho$ of $C$ using the same variable ordering of the path in the BDD. This notion is captured in the following lemma.

LEMMA 6.7 *Given a BDD $F$ and a CSAT formula $\rho$ for some logic circuit $C$, then a truth assignment $A_p$ on a certain path $p$ from the root of $F$ to the terminal, $\rho$ is satisfiable using the same variable ordering and truth assignment.*

**Proof.** Assume the set of primary inputs is $X_n$. Then we can partition $X_n$ into two sets, $S$ and $T$, such that $S$ contains all the variables that correspond to nodes in $p$ and $T$ contains all the remaining variables. By the Deletion Rule from Section 2.4.3, $T$ is the set of all redundant nodes. But due to the equivalence of the Deletion Rule and Definition 6.3, it follows from Theorem 6.4 that $\rho$ is satisfied. □

From this perspective, the equivalence checking problem between two circuits can be viewed as a search in the decision trees of the two circuits for a path that leads to the terminal node **1** (**0**) in one but leads to the terminal node **0** (**1**) in the other. This view allows the introduction of the relation between BDDs and the DP procedure as given by the following theorem.

THEOREM 6.8 *Given a BDD $F$ with a number of paths $P$ and a CCNF formula $\rho$ for some logic circuit $C$ then if the variable ordering strategy of the DP procedure follows the same ordering for every path of $F$, then DP proves the equivalence of $C$ against an equivalent version in $P - 1$ backtracks.*

**Proof.** In order to prove the equivalence of two circuits, we have to check that every assignment that leads to the terminal node **1** (**0**) in the BDD of one circuit leads to the same result in the other circuit. Consequently, if the variable ordering for splitting in the DP procedure is the same as in the corresponding BDD, then considering an alternative path in the BDD leads to a backtrack in the DP procedure. Thus, the number of paths exceeds the total number of backtracks exactly by one. □

We now utilize Theorem 6.8 to calculate optimal lower bounds on the number of backtracks that can be obtained from the DP procedure using the primary inputs for backtracking.

Searching among the $n!$ different variable orderings for a BDD over $X_n$ helps identifying the variable ordering(s) that produces the minimal number of paths. This minimal number of paths allows the calculation of a lower bound on the number of backtracks in the DP procedure as given by the following theorem.

THEOREM 6.9 *Given a DP procedure that operates using Theorem 6.4 and 6.5, then the optimal number of backtracks needed to prove the equivalence of two equivalent circuits is bounded by the number of paths in the corresponding minimal path BDD.*

**Proof.** Since the number of paths and backtracks are linked by Theorem 6.8 and there exists a variable ordering that minimizes the number of paths, then the optimal number of backtracks using the DP procedure can be obtained if we follow the same variable ordering for every path in the minimal path BDD. □

The importance of Theorem 6.9 is that it allows benchmarking any static variable ordering strategy for the DP procedure against an optimal lower bound. We now develop a variable ordering heuristic for the DP procedure that tries to trace the same variable ordering for every path in the corresponding minimal path BDD of the circuit.

## 6.3 Dynamic Variable Ordering Strategy for DP Procedure

From the previous section, we conclude that the variable ordering strategy should differ for every path of the decision tree, and furthermore result in no splitting on a redundant variable under the current partial assignment. In this section, we propose a structural method that avoids redundant splitting and tries to make the fewest possible splittings to satisfy $\rho$. The method is based on the following theorem.

THEOREM 6.10 *If a bounded gate lies on every path from the primary output $z$ to the unassigned primary input $\nu \in (X_n \setminus S)$ under a current partial assignment $A_S$ for a set $S \subset X_n$, then $f_{A_S}|_{\nu=0} = f_{A_S}|_{\nu=1}$.*

**Proof.** The existence of a bounded gate (a gate with a specified output) in every path to $z$ implies that no value assigned to $\nu$ can affect the function of the circuit since these bounded gates will suppress the propagation of the logic value. Thus, since $f(A_S)$ remains unchanged under any assignment to $\nu$, $f_{A_S}|_{\nu=0} = f_{A_S}|_{\nu=1}$. □

In order to utilize Theorem 6.10, a DP variable ordering strategy was proposed that is a modification of a BDD variable ordering heuristic by [Min96]. The method starts by assigning a weight of 1.0 to the primary output, and continues by propagating this weight to the primary inputs in the following manner: divide the output weight of each gate among its inputs, accumulate the weight of the fan-out branches into the fan-out stem. Next, the primary input with the highest weight is chosen as the next variable to be split in the DP procedure. By unit clause propagation, we exclude all the gate outputs of the bounded literals from the next cycle of weight calculations.

The proposed weight assignment method assigns a weight of zero to every redundant primary input under the current partial assignment. Furthermore, the strategy splits on the largest weight input in an effort to obtain a minimal number of splittings for satisfying the CSAT formula. We illustrate the weight calculation procedure by the following example.

EXAMPLE 6.11 *Suppose we are proving the circuit shown in Figure 6.3(a) against an equivalent version of it. In this case we have to compare if their BDDs are isomorphic or using the DP procedure, we have to check that every path results in the same output assignment in both circuits. Applying the previous procedure of weight calculations to minimize the number of paths being compared, we trace the different paths of the BDD of the circuit in Figure 6.3(a) in the manner shown below.*

*From the initial weights given in Figure 6.3(a), we split on $x_3$ since it has the highest weight. Suppose $x_3$ is assigned the value 1. Then by unit clause propagation, gate h becomes bounded to the truth value 0. The weight calculation procedure is executed again but this time by setting the weights of bounded gates to zero. Figure 6.3(a) illustrates the new weights. Next, $x_1, x_2$ or $x_3$ can be chosen to split on; the resultant search tree will always produce identical number of backtracks. Notice that bounded gates are marked with a "*" in the corresponding figure. After finishing this half of the decision tree, $x_3$ is flipped and assigned to 0. The new calculated weights are shown in Figure 6.3(d). Since $x_4$ has the highest weight, we split on $x_4$ thus reaching the terminals.*

*We now construct the minimum path BDD corresponding to this circuit as shown in Figure 6.3(d). We notice in the BDD that the splitting choices made by the previous procedure lead to traversing this BDD in*

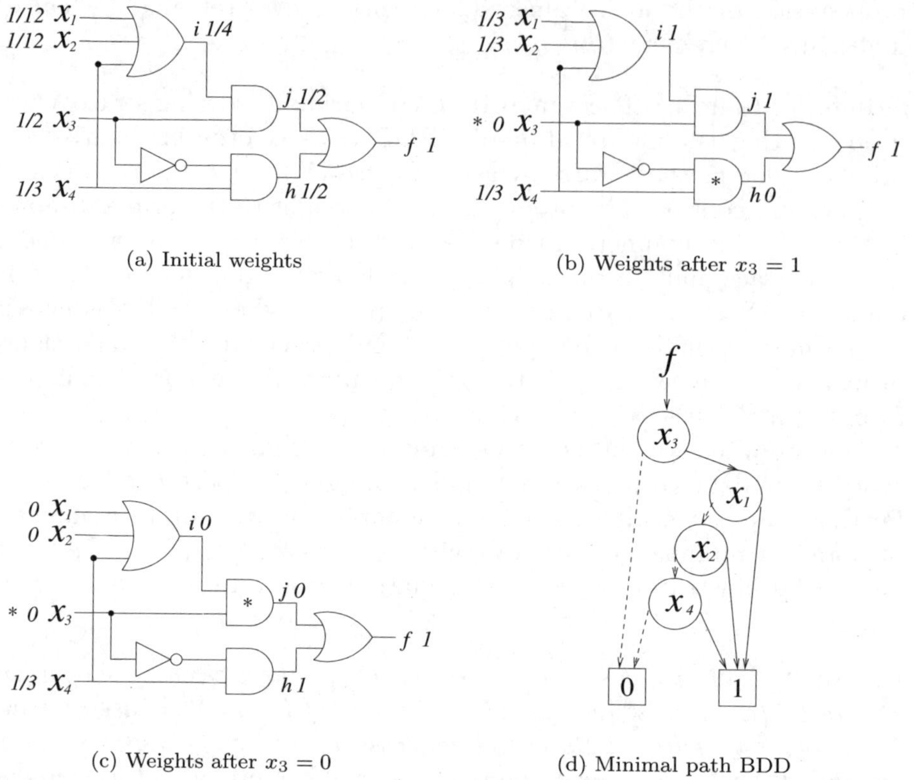


(a) Initial weights

(b) Weights after $x_3 = 1$

(c) Weights after $x_3 = 0$

(d) Minimal path BDD

*Figure 6.3.* Circuit from Example 6.11.

*exact ordering for each path, thus producing the minimal number of backtracks, 5, corresponding to the minimal number of paths, 6.*

As was just illustrated, the proposed method has produced optimal results for the proposed example; however, the method is not in general optimal since it depends on the structure of the circuit. But as shall be demonstrated in the next section, the proposed method is capable of achieving near optimal results.

## 6.4 Experimental Results

In this section, we present experimental results for our proposed approach. The experiments have been carried out on a PC with an Intel Pentium 233 MHz processor and 64MB of physical memory. TEGUS

*Table 6.1.* Comparing greedy search to BDD-based ordering for the DP procedure

| name | output | Greedy Search | | BDD-based | | Difference | | BDD |
|---|---|---|---|---|---|---|---|---|
| | | backt. | time | backt. | time | backt. | time | #paths |
| c0432 | 370gat | Abort | 771 | 943,200 | 185 | 95.3% | 76.0% | 894,606 |
| | 432gat | Abort | 770 | 816,012 | 235 | 95.9% | 69.5% | 798,906 |
| | 329gat | $1.6 \cdot 10^6$ | 22.50 | 81,912 | 5.94 | 95.0% | 73.6% | 57,513 |
| c0499 | od1 | Abort | 1,686 | Abort | 10,351 | - | - | $20 \cdot 10^6$ |
| | od18 | Abort | 1,685 | Abort | 10,245 | - | - | $23 \cdot 10^6$ |
| | od31 | Abort | 1,691 | Abort | 10,627 | - | - | $23 \cdot 10^6$ |
| c0880 | 864 | 649,741 | 12.30 | 164,656 | 27.50 | 74.7% | -123.6% | 89,937 |
| | 850 | 37,225 | 0.60 | 24,171 | 3.45 | 35.1% | -475.0% | 21,981 |
| | 874 | Abort | 498 | $1.1 \cdot 10^6$ | 278 | 94.6% | 44.2% | 532,793 |
| c1355 | 1353 | Abort | 1,921 | Abort | 8,501 | - | - | $21 \cdot 10^6$ |
| | 1354 | Abort | 1,892 | Abort | 8,427 | - | - | $23 \cdot 10^6$ |
| | 1355 | Abort | 1,984 | Abort | 8,602 | - | - | $23 \cdot 10^6$ |
| c1908 | 57 | Abort | 718 | 147,457 | 11.90 | 99.3% | 98.3% | 134,221 |
| | 60 | Abort | 1,708 | 28,161 | 20.10 | 99.9% | 98.8% | 26,638 |
| | 66 | Abort | 2,058 | 14,249 | 10.40 | 99.9% | 99.5% | 12,638 |
| c3540 | 399 | 24,550 | 0.73 | 4,528 | 1.34 | 81.6% | -83.6% | 1,636 |
| | 384 | $6.1 \cdot 10^6$ | 652 | 70997 | 57.20 | 98.8% | 91.2% | 65,535 |
| | 387 | $1.6 \cdot 10^6$ | 213 | 17312 | 17 | 98.9% | 92.0% | 16,456 |
| C5315 | 688 | Abort | 300 | 23974 | 13 | 99.9% | 95.7% | 13,919 |
| | 843 | Abort | 3383 | 211,181 | 162 | 98.9% | 95.2% | 103,569 |
| | 818 | 85,989 | 3.50 | 6537 | 2.18 | 92.4% | 37.7% | 2,489 |
| C7522 | 373 | 16 | 0.01 | 10 | 0.01 | 37.5% | 0% | 10 |
| | 376 | 40,952 | 1.30 | 15416 | 4.93 | 62.4% | -279.2% | 7,152 |
| | 359 | $2.5 \cdot 10^6$ | 114 | 585,584 | 285 | 76.3% | -150.0% | 196,592 |
| c6288 | 2223 | 157 | 0.01 | 102 | 0.01 | 35.0% | 0% | 102 |
| | 3895 | 174,315 | 12.20 | 106,423 | 108 | 38.9% | -785.3% | 109,668 |
| | 4591 | $2.8 \cdot 10^6$ | 255 | $1.7 \cdot 10^6$ | 2,689 | 39.0% | -954.5% | $1.8 \cdot 10^6$ |

[SBSV92] was used as a SAT solver and CUDD [Som02] as decision diagram package.

In the experiments the equivalence check of the ISCAS85 benchmark circuits [BF85] against their non-redundant version was considered. The proposed approach was benmarked against the greedy search approach of TEGUS. Table 1 gives the results of such a comparison. The first column lists the circuit name. In the second column, the name of the output under verification is provided. The next two columns give the number of backtracks and time needed to prove the output using the TEGUS approach. In the next two columns, the number of backtracks and time needed to prove the output using the proposed approach are given. The possible "Abort" notation in the third and the fifth column denotes reaching the 20 million backtrack threshold without reaching a successful resolution; the subsequent time columns in that case denote the time needed to reach this limit. Regarding the last three columns, the first two columns report the percentage decrease in the number of backtracks and time, respectively. The last column provides the near minimal number of paths in the corresponding BDD. In obtaining the number of paths, we only consider a subspace of the possible variable orderings. We compare the number of paths that result from reducing the corresponding BDD to a minimum size by sifting (Section 2.4.7.2) and picking the minimal path. Three outputs are considered for each circuit.

Comparing the greedy approach to the proposed variable ordering strategy, we observe that the proposed approach results on the average in 90% decrease in the number of backtracks. The time needed to complete the backtracks varies from reduction in 13 cases with an average of 70% decrease to increases in about 7 cases with an average increase of about 4 times. While at first glance the time results look inconclusive, it can easily be noted that the increases are essentially associated with low-backtrack, provable outputs, since in these cases the recurring weight assignment costs cannot be amortized across the small number of backtracks. The approach improves the performance of large search tree cases, the main focus of practical interest. We also notice that within the limit of 20 million backtracks, TEGUS fails to complete on 7 outputs while the proposed approach proves them in a relatively low number of backtracks.

As proved in Section 3, the minimum number of paths that can be obtained from the corresponding BDD is a lower bound on the number of backtracks that can be obtained in the DP procedure. The number of paths provides an ability to benchmark various ordering techniques and also provides an insight on the performance of the DP procedure. For example, the huge number of paths in case of the c0499 and c1355 circuits (they are functionally equivalent) suggests that they are hard-to-prove outputs using the DP procedure. In addition, in circuits, like the c6288, where the variation of the number of paths with respect to the variable ordering is small, the dynamic weight assignment technique constitutes a time consuming step with respect to the overall time, as the number of backtracks varies slightly with respect to the variable splitting strategy.

## 6.5 Summary

In this chapter the relation between the search tree of the DP procedure and the BDD of the corresponding function has been studied. The direct relation between the number of paths in a BDD and the number of backtracks needed to prove the equivalence of two functionally equivalent circuits has been established. This relation introduces the ability to calculate an optimal lower bound on the number of backtracks needed to prove the equivalence checking problem. In addition, this relation has led to the conclusion that the capture of the variable ordering of the minimal path BDD in the DP procedure implies a reduction in the number of backtracks needed to prove the problem. In order to exploit this relation, we have devised a variable ordering technique for the DP procedure that through experimental results has exhibited superior performance in terms of the number of backtracks and time needed to prove the equivalence problem. The relation between the two procedures offers novel ways of tight integration of different prover approaches in a verification tool.

Chapter 7

# FINAL REMARKS

During the last twenty years, a large number of problems in VLSI CAD have arisen that can be tackled by the use of *Binary Decision Diagrams* (BDDs). Besides this, also the clausal representation as a *Boolean Satisfiability* (SAT) problem has been frequently used. The applications are in various fields including logic synthesis, formal and simulation-based verification, and design-for-testability. Facing the ever increasing design complexity and the growing pressure of time-to-market, new approaches for design space exploration should be considered. Moving towards new optimization goals, new efficient algorithms for these tasks are needed.

Recent methods utilizing BDDs require the optimization of BDDs with respect to new, alternative objective functions. In addition, the lastest trends in research move towards the *fusion* of SAT and BDD as well-established concepts. However, the relation between the two paradigms and the basic concepts of new optimization goals have been known and understood for a short time.

At this point it should not be overseen that *Artificial Intelligence* (AI) can be a source of new impulses. In the past, a number of methods in VLSI CAD have been based on concepts that originally have been developed in the AI community, e.g. *Evolutionary Algorithms* (EA) and *Simulated Annealing* (SA). Basic search techniques like *hill climbing* have been reflected by classical BDD techniques like the sifting algorithm for dynamic BDD reordering. The latest proposal is to use the paradigm of the generic $A^*$*-algorithm* to reduce the run time of tasks in VLSI CAD that are of time-limited nature.

In this book, the classical and the latest approaches to the optimization of BDDs with respect to the classical criterion, i.e. the size of the diagram, and new, alternative criteria have been presented. Recent methods can yield reductions in run time of orders of magnitude compared to classical algorithms. Reflecting a second important trend, the relation between BDD and SAT has been studied considering formal equivalence checking. The presented material covers the recent developments and also includes the established methods.

An important development shows a growing number of methods that require BDD optimization or work towards a fusion of the concepts BDD and SAT. Some of them achieve improvements by using classical techniques of AI. Even though still in its early stage, the respective paradigm shift in VLSI CAD already shows a promising potential. New algorithms have to be developed that cope with challenging design spaces and emerging optimization objectives. This especially holds in the presence of the increasing design gap.

# References


[AHU74] A.V. Aho, J.E. Hopcroft, and J.D. Ullman. *The Design and Analysis of Computer Algorithms.* Addison–Wesley, 1974.

[Ake78a] S.B. Akers. Binary decision diagrams. *IEEE Trans. on Comp.*, 27:509–516, 1978.

[Ake78b] S.B. Akers. Functional testing with binary decision diagrams. In *Eighth Annual Conference on Fault-Tolerant Computing*, pages 75–82, 1978.

[AM95] P. Ashar and S. Malik. Fast functional simulation using branching programs. In *Int'l Conf. on CAD*, pages 408–412, 1995.

[BCCZ99] A. Biere, A. Cimatti, E. Clarke, and Y. Zhu. Symbolic model checking without BDDs. In *Tools and Algorithms for the Construction and Analysis of Systems*, volume 1579 of *Lecture Notes in Computer Science.* Springer Verlag, 1999.

[Bec92] B. Becker. Synthesis for testability: Binary decision diagrams. In *Symp. on Theoretical Aspects of Comp. Science*, volume 577 of *LLNCS*, pages 501–512. Springer Verlag, 1992.

[Bec98] B. Becker. Testing with decision diagrams. *INTEGRATION, the VLSI Jour.*, 26:5–20, 1998.

[BF85] F. Brglez and H. Fujiwara. A neutral netlist of 10 combinational circuits and a target translator in fortran. In *Int'l Symp. Circ. and Systems, Special Sess. on ATPG and Fault Simulation*, pages 663–698, 1985.

[BLW95] B. Bollig, M. Löbbing, and I. Wegener. Simulated annealing to improve variable orderings for OBDDs. In *Int'l Workshop on Logic Synth.*, pages 5b:5.1–5.10, 1995.

[BLW96] B. Bollig, M. Löbbing, and I. Wegener. On the effect of local changes in the variable ordering of ordered decision diagrams. *Information Processing Letters*, 59:233–239, 1996.

[BMP97] L. Benini, E. Macii, and M. Poncino. Telescopic units: Increasing the average throughput of pipelined designs by adaptive latency control. In *Design Automation Conf.*, pages 22–27, 1997.

[BNNSV97] P. Buch, A. Narayan, A. Newton, and A. Sangiovanni-Vincentelli. Logic synthesis for large pass transistor circuits. In *Int'l Conf. on CAD*, pages 663–670, 1997.

[BRB90] K. Brace, R. Rudell, and R.E. Bryant. Efficient implementation of a BDD package. In *Design Automation Conf.*, pages 40–45, 1990.

[Bry86] R. E. Bryant. Graph-based algorithms for Boolean function manipulation. *IEEE Trans. on Comp.*, 35(8):677–691, 1986.

[Bry91] R. E. Bryant. On the complexity of VLSI implementations and graph representations of Boolean functions with application to integer multiplication. *IEEE Trans. on Comp.*, 40:205–213, 1991.

[BS98] J.R. Burch and V. Singhal. Tight integration of combinational verification methods. In *Int'l Conf. on CAD*, pages 570–576, 1998.

[BW96] B. Bollig and I. Wegener. Improving the variable ordering of OBDDs in NP-complete. *IEEE Trans. on Comp.*, 45(9):993–1002, 1996.

[CG85] E. Cerny and J. Gecsei. Simulation of mos circuits by decision diagrams. *IEEE Trans. on CAD*, CAD-4(4):685–693, 1985.

[CLAB98] R. Chaudhry, T.-H. Liu, A. Aziz, and J.L. Burns. Area-oriented synthesis for pass-transistor logic. In *Int'l Conf. on Comp. Design*, pages 160–167, 1998.

[CNQ03] G. Cabodi, S. Nocco, and S. Quer. SAT-based bounded model checking by means of BDD-based approximate traversals. In *Design, Automation and Test in Europe*, pages 898–903, 2003.

[Col93] Collaborative Benchmarking Laboratory. *1993 LGSynth Benchmarks.* North Carolina State University, Department of Computer Science, 1993.

[DB98] R. Drechsler and B. Becker. *Binary Decision Diagrams - Theory and Implementation.* Kluwer Academic Publishers, 1998.

[DDG00] R. Drechsler, N. Drechsler, and W. Günther. Fast exact minimization of BDDs. *IEEE Trans. on CAD*, 19(3):384–389, 2000.

[DG97] R. Drechsler and N. Göckel. Minimization of BDDs by evolutionary algorithms. In *Int'l Workshop on Logic Synth.*, 1997.

[DG02] R. Drechsler and W. Günther. *Towards One-Pass Synthesis.* Kluwer Academic Publishers, 2002.

[DGB96] R. Drechsler, N. Göckel, and B. Becker. *Parallel Problem Solving from Nature*, volume 577 of *LNCS*, chapter Learning heuristics for OBDD minimization by evolutionary algorithms, pages 730–739. Springer Verlag, 1996.

[DGS01] R. Drechsler, W. Günther, and F. Somenzi. Using lower bounds during dynamic BDD minimization. *IEEE Trans. on CAD*, 20(1):51–57, 2001.

[DLL62] M. Davis, G. Logeman, and D. Loveland. A machine program for theorem proving. *Communications of the ACM*, 5:394–397, 1962.

[DM99] E.V. Dubrova and D.M. Miller. On disjoint covers and ROBDD size. In *Pacific Rim Conference on Communications, Computers and Signal Processing*, pages 162–164, 1999.

[DP60] M. Davis and H. Putnam. A computing procedure for quantification theory. *Journal of the ACM*, 7:506–521, 1960.

[DP87] R. Dechter and J. Pearl. Generalized best-first search strategies and the optimality of $A^*$. *Journal of the Association for Computing Machinery*, 34(1):1–38, 1987.

[DSF04] R. Drechsler, J. Shi, and G. Fey. Synthesis of fully testable circuits from BDDs. *IEEE Trans. on CAD*, 23(3):440–443, 2004.

[Ebe03] R. Ebendt. Reducing the number of variable movements in exact BDD minimization. In *Int'l Symp. on Circuits and Systems*, volume 5, pages 605–608, 2003.

[ED05] R. Ebendt and R. Drechsler. Lower bounds for dynamic BDD reordering. In *Asia and South Pacific Design Automation Conf.*, pages 579–582, 2005.

[EGD03a] R. Ebendt, W. Günther, and R. Drechsler. Combination of lower bounds in exact BDD minimization. In *Design, Automation and Test in Europe*, pages 758–763, 2003.

[EGD03b] R. Ebendt, W. Günther, and R. Drechsler. An improved branch and bound algorithm for exact BDD minimization. *IEEE Trans. on CAD*, 22(12):1657–1663, 2003.

[EGD04a] R. Ebendt, W. Günther, and R. Drechsler. Combining ordered best-first search with branch and bound for exact BDD minimization. In *Asian and South Pacific Design Automation Conf.*, pages 876–879, 2004.

[EGD04b] R. Ebendt, W. Günther, and R. Drechsler. Minimization of the expected path length in BDDs based on local changes. In *Asian and South Pacific Design Automation Conf.*, pages 866–871, 2004.

[EGD05] R. Ebendt, W. Günther, and R. Drechsler. Combining ordered-best first search with branch and bound for exact BDD minimization. *IEEE Trans. on CAD*, 2005.

[Fal93] B.J. Falkowski. Calculation of Rademacher-Walsh spectral coefficients for systems of completely and incompletely specified boolean functions. In *IEEE Proceedings on Circuits*, pages 1698–1701, 1993.

[FC99] B.J. Falkowski and C.-H. Chang. Paired Haar spectra computation through operations on disjoint cubes. In *IEEE Proceedings on Circuits, Devices and Systems*, pages 117–123, 1999.

[FD02a] G. Fey and R. Drechsler. Minimizing the number of paths in BDDs. In *15th Symp. on Integrated Circuits and System Design*, pages 359–364, 2002.

[FD02b] G. Fey and R. Drechsler. Utilizing BDDs for disjoint SOP minimization. In *45th IEEE International Midwest Symp. on Circuits and Systems*, pages 306–309, 2002.

[FMK91] M. Fujita, Y. Matsunaga, and T. Kakuda. On variable ordering of binary decision diagrams for the application of multilevel synthesis. In *European Conf. on Design Automation*, pages 50–54, 1991.

[FMM+98] F. Ferrandi, A. Macii, E. Macii, M. Poncino, R. Scarsi, and F. Somenzi. Symbolic algorithms for layout-oriented synthesis of pass transistor logic circuits. In *Int'l Conf. on CAD*, pages 235–241, 1998.

[FOH93] H. Fujii, G. Ootomo, and C. Hori. Interleaving based variable ordering methods for ordered binary decision diagrams. In *Int'l Conf. on CAD*, pages 38–41, 1993.

[FS90] S.J. Friedman and K.J. Supowit. Finding the optimal variable ordering for binary decision diagrams. *IEEE Trans. on Comp.*, 39(5):710–713, 1990.

[FSC93] B.J. Falkowski, I. Schäfer, and C.-H. Chang. An effective computer algorithm for the calculation of disjoint cube representation of boolean functions. In *Midwest Symposium on Circuits and Systems*, pages 1308–1311, 1993.

[FSD04] G. Fey, J. Shi, and R. Drechsler. BDD circuit optimization for path delay fault testability. In *EUROMICRO Symp. on Digital System Design 2004*, pages 162–172, 2004.

[GD00] W. Günther and R. Drechsler. Improving EAs for sequencing problems. In *Genetic and Evolutionary Computation Conf.*, pages 175–180, 2000.

[GGW+03] A. Gupta, M. Ganai, C. Wang, Z. Yang, and P. Ashar. Abstraction and BDDs complement SAT-based BMC in DiVer. In *International Conference on Computer Aided Verification*, volume 2725 of *Lecture Notes in Computer Science*, pages 206–209. Springer-Verlag, 2003.

[GN02] E. Goldberg and Y. Novikov. BerkMin: a fast and robust SAT-solver. In *Design, Automation and Test in Europe*, pages 142–149, 2002.

[Goe81] P. Goel. An implicit enumeration algorithm to generate test for combinational logic. *IEEE Trans. on Comp.*, 30:215–222, 1981.

[GYA+01] A. Gupta, Z. Yang, P. Ashar, L. Zhang, and S. Malik. Partition-based decision heuristics for image computation using SAT and BDDs. In *Int'l Conf. on CAD*, pages 286–292, 2001.

[HNR68] P.E. Hart, N.J. Nilsson, and B. Raphael. A formal basis for the heuristic determination of minimum cost paths. *IEEE Trans. Syst. Sci. Cybern.*, 2:100–107, 1968.

[ISM03] Y. Iguchi, T. Sasao, and M. Matsuura. Evaluation of multiple-output logic functions using decision diagrams. In *Asian South Pacific Design Automation Conf.*, pages 312–315, 2003.

[ISY91] N. Ishiura, H. Sawada, and S. Yajima. Minimization of binary decision diagrams based on exchange of variables. In *Int'l Conf. on CAD*, pages 472–475, 1991.

[JKCMS97] Y.-M. Jiang, A. Krstic, K.-T. Cheng, and M. Marek-Sadowska. Post-layout logic restructuring for performance optimization. In *Design Automation Conf.*, pages 662–665, 1997.

[JKS93] S.-W. Jeong, T.-S. Kim, and F. Somenzi. An efficient method for optimal BDD ordering computation. In *Int'l Conf. on VLSI and CAD*, 1993.

[JMB03] Y. Jiang, S. Matic, and R.K. Brayton. Generalized cofactoring for logic function evaluation. In *Design Automation Conf.*, pages 155–158, 2003.

[Kar88] K. Karplus. Representing boolean functions with if-then-else dags. Computer Engineering UCSC-CRL-88-28, 1988.

[KRR88] V. Kumar, K. Ramesh, and V.N. Rao. Parallel best-first search of state-space graphs: A summary of results. In *National Conf. on Artificial Intelligence*, pages 122–127, 1988.

[KS95] K. Kuroda and T. Sakurai. Overview of low-power ulsi circuit techniques. *IEICE Trans. on Information and Systems*, E78-C(4):334–343, 1995.

[LAB98] T. H. Liu, A. Aziz, and J.L. Burns. Performance driven synthesis for pass-transistor logic. In *Int'l Workshop on Logic Synth.*, pages 255–259, 1998.

[Lar92] T. Larrabee. Test pattern generation using Boolean satisfiability. *IEEE Trans. on CAD*, 11:4–15, 1992.

[Lee59] C.Y. Lee. Representation of switching circuits by binary decision diagrams. *Bell System Technical Jour.*, 38:985–999, 1959.

[LWA98] Y. Luo, T. Wongsonegoro, and A. Aziz. Hybrid techniques for fast functional simulation. In *Design Automation Conf.*, pages 664–667, 1998.

[LWHL01] Y. Y. Liu, K. H. Wang, T. T. Hwang, and C. L. Liu. Binary decision diagram with minimum expected path length. In *Design, Automation and Test in Europe*, pages 708–712, 2001.

[Mar77] A. Martelli. On the complexity of admissable search algorithms. *Artificial Intelligence*, 23:1–13, 1977.

[MB88] J.-C. Madre and J.-P. Billon. Proving circuit correctness using formal comparison between expected and extracted behavior. In *Design Automation Conf.*, pages 205–210, 1988.

[MBM01] L. Macchiarulo, L. Benini, and E. Macii. On-the-fly layout generation for PTL macrocells. In *Design, Automation and Test in Europe*, pages 546–551, 2001.

[MBSV95] R. Murgai, R.K. Brayton, and A.L. Sangiovanni-Vincentelli. *Logic Synthesis for Field-Programmable Gate Arrays.* Kluwer Academic Publishers, 1995.

[Min96] S. Minato. *Binary Decision Diagrams and applications for VLSI CAD.* Kluwer Academic Publisher, 1996.

[MIY90] S. Minato, N. Ishiura, and S. Yajima. Shared binary decision diagram with attributed edges for efficient function manipulation. In *Design Automation Conf.*, pages 52–57, 1990.

[MMS+95] P.C. McGeer, K.L. McMillan, A. Saldanha, A.L. Sangiovanni-Vincentelli, and P. Scaglia. Fast discrete function evaluation using decision diagrams. In *Int'l Conf. on CAD*, pages 402–407, 1995.

[MMZ+01] M.W. Moskewicz, C.F. Madigan, Y. Zhao, L. Zhang, and S. Malik. Chaff: Engineering an efficient SAT solver. In *Design Automation Conf.*, pages 530–535, 2001.

[Mo84] L. Mérõ. A heuristic search algorithm with modifiable estimate. *Artificial Intelligence*, 23:13–27, 1984.

[MP01] A. Mishchenko and M. Perkowski. Fast heuristic minimization of exclusive-sums-of-products. In *Int'l Workshop on Applications of the Reed-Muller Expansion in Circuit Design*, pages 242–250, 2001.

[MS97] C. Meinel and A. Slobodová. Speeding up variable reordering of OBDD. In *Int'l Conf. on Comp. Design*, pages 338–343, 1997.

[MS98] J.P. Marques-Silva. An overview of backtrack search satisfiability algorithms. In *International Symposium on AI and Mathematics*, 1998.

[MSG99] J.P. Marques-Silva and T. Glass. Combinational equivalence checking using boolean satisfiability and recursive learning. In *Design, Automation and Test in Europe*, pages 145–149, 1999.

[MSMSL99] A. Mukherjee, R. Sudhakar, M. Marek-Sadowska, and S. Long. Wave steering in YADDs: a novel, non-iterative synthesis and layout technique. In *Design Automation Conf.*, pages 466–471, 1999.

[MSS96] J.P. Marques-Silva and K.A. Sakallah. GRASP – a new search algorithm for satisfiability. In *Int'l Conf. on CAD*, pages 220–227, 1996.

[Nil80] N. Nilsson. *Principles of Artificial Intelligence.* Tioga Publishing Company, Palo Alto, CA, 1980.

[NMSB03] S. Nagayama, A. Mishchenko, T. Sasao, and J. Butler. Minimization of average path length in BDDs by variable reordering. In *Proc. of International Workshop on Logic and Synthesis*, 2003.

[PCK99] M. Prasad, P. Chong, and K. Keutzer. Why is ATPG easy? In *Design Automation Conf.*, pages 22–28, 1999.

[Pea84] J. Pearl. *Heuristics: Intelligent Search Strategies for Computer Problem Solving.* Addison-Wesley, 1984.

[PK00] V. Paruthi and A. Kuehlmann. Equivalence checking combining a structural SAT-solver, BDDs, and simulation. In *Int'l Conf. on Comp. Design*, pages 459–464, 2000.

[PS95] S. Panda and F. Somenzi. Who are the variables in your neighbourhood. In *Int'l Conf. on CAD*, pages 74–77, 1995.

[RDO02] S. Reda, R. Drechsler, and A. Orailoglu. On the relation between SAT and BDDs for equivalence checking. In *Int'l Symp. on Quality of Electronic Design*, pages 394–399, 2002.

[Ric88] E. Rich. *Artificial Intelligence.* McGraw–Hill, 1988.

[RMSS98] K. Ravi, K.L. McMillan, T.R. Shiple, and F. Somenzi. Approximation and decomposition of binary decision diagrams. In *Design Automation Conf.*, pages 445–450, 1998.

[RS95] K. Ravi and F. Somenzi. High-density reachability analysis. In *Int'l Conf. on CAD*, pages 154–158, 1995.

[RS01] S. Reda and A. Salem. Combinational equivalence checking using binary decision diagrams and Boolean satisfiability. In *Design, Automation and Test in Europe*, pages 122–126, 2001.

[Rud93] R. Rudell. Dynamic variable ordering for ordered binary decision diagrams. In *Int'l Conf. on CAD*, pages 42–47, 1993.

[Sas93] T. Sasao. EXMIN2: A simplification algorithm for Exclusive-OR-Sum-of products expressions for multiple-valued-input two-valued-output functions. *IEEE Trans. on CAD*, 12:621–632, 1993.

[SB00] C. Scholl and B. Becker. On the generation of multiplexer circuits for pass transistor logic. In *Design, Automation and Test in Europe*, pages 372–378, 2000.

[SBSV92] P. R. Stephan, R. K. Brayton, and A. Sangiovanni-Vincentelli. Combinational test generation using satisfiability. Technical Report

UCB/ERL M92/112, Dept. of EECS, Univ. of California, Berkeley, October 1992.

[SDB97] C. Scholl, R. Drechsler, and B. Becker. Functional simulation using binary decision diagrams. In *Int'l Conf. on CAD*, pages 8–12, 1997.

[Sha38] C.E. Shannon. A symbolic analysis of relay and switching circuits. *Trans. AIEE*, 57:713–723, 1938.

[SM98] A. Slobodová and C. Meinel. Sample method for minimization of OBDD. In *Int'l Workshop on Logic Synth.*, pages 311–316, 1998.

[Som01] F. Somenzi. Efficient manipulation of decision diagrams. *Software Tools for Technology Transfer*, 3(2):171–181, 2001.

[Som02] F. Somenzi. *CU Decision Diagram Package Release 2.3.1.* University of Colorado at Boulder, 2002.

[ST02] L. Shivakumaraiah and M. Thornton. Computation of disjoint cube representations using a maximal binate variable heuristic. In *Southeastern Symposium on System Theory*, pages 417–421, 2002.

[TDM01] M. Thornton, R. Drechsler, and D.M. Miller. *Spectral Techniques in VLSI CAD.* Kluwer Academic Publisher, 2001.

[YC02] C. Yang and M.J. Ciesielski. BDS: a BDD-based logic optimization system. *IEEE Trans. on CAD*, 21(7):866–876, 2002.

[YSRS96] K. Yano, Y. Sasaki, K. Rikino, and K. Seki. Top-down pass transistor logic design. *IEEE Jour. of Solid-State Circ.*, 31(6):792–803, 1996.

[ZA01] H. Zhou and A. Aziz. Buffer minimization in pass transistor logic. *IEEE Trans. on CAD*, 20(5):693–697, 2001.

# Index